PRAISE FOR
Finding the Elephant

—§—

'Finding the Elephant' is a clever synthesis of ideas from contemporary physics, biology, psychology and religion, all converging toward the idea of a deeper reality underpinning our everyday experience. Delightfully well-written, persuasive, and up-to-date.

—Dean Radin, PhD, Chief Scientist at the Institute of Noetic Sciences (IONS) and Volunteer Faculty in the Department of Psychology at Sonoma State University, Co-Editor-in-Chief of the Journal *Explore* and award-winning and best-selling author: *The Conscious Universe*, *Entangled Minds* and *Supernormal*.

—§—

Here we are living in hyperspecialized modern society that is producing a cornucopia of technological wonders. But in the process the people enjoying these toys have almost totally forgotten who they are as human beings connected with themselves and the universe. David Burfoot's book proposes we find the elephant, the larger context of consciousness and our place in the universe instead of focusing on the material plane that the majority of contemporary scientists want us to believe is the total elephant. He provides a plethora of perspectives to show the extreme limitations of the current scientific/religious materialist perspective that modern western education inculcates so well.

—Arthur Buehler, PhD, Senior Lecturer, School of Art History, Classics and Religious Studies, Victoria University of Wellington, New Zealand

—§—

This book is a very welcome addition to the growing literature on post-materialist science. By reading this update, the reader can only conclude that a consciousness-based paradigm is aborning.

—Amit Goswami, PhD, retired full Professor from the University of Oregon's Department of Physics, U.S.A., and author of *Quantum Creativity: Think Quantum, Be Creative* and *Quantum Economics: Unleashing the Power of an Economics of Consciousness.*

FINDING THE ELEPHANT

FINDING THE ELEPHANT®

WWW.FINDINGTHEELEPHANT.NET

Copyright © 2017 by David Burfoot

All rights reserved. This book or any portion thereof may not be reproduced or used in any manner whatsoever without the express written permission of the publisher and author except for the use of brief quotations in a book review.

The images used in this book are original, used with the permission of the owner (or in line with the Creative Commons licence) or considered Public Domain. If an owner of any of the used images has an objection to their use, they may contact the author through the below website.

Printed in Australia

First Printing, 2017

ISBN 978-0-9944859-0-8 (Paperback)

ISBN 978-0-9944859-1-5 (Electronic book)

www.findingtheelephant.net

Subspace, the Mega-phenomenon

Dedicated to Jacquie, Kaemon and Kayode
and
in loving memory of Stella, Nigel and Adrian

TABLE OF CONTENTS

Acknowledgements i

Introduction iii

Chapter 1 1

Space, the Final Frontier 3

Six Blind Men and the Elephant 6

Triangulation 6

    Validation and complementary information 9

Group intelligence research 10

The invisible elephant 12

It's not their fault 21

Beware first principles 28

This 'is science' and this is 'not science' 30

Summary 32

Dissecting the elephant 34

Chapter 2 35

Non-material reality—the natural forces 37

Quantum mechanics 39

Nature of matter 40

Matter is an emergent property 42

Entanglement 44

Space-time 46

    Past, present and future continuum 47

    Why now? 48

Finding the elephant in string theory and wormholes. 50

    Wormholes, time warps and warp drives 51

Light and the role of the observer 53

    Delayed-choice experiments 57

Anthropic principle 62

The universe—as a quantum state 66

Summary 69

Chapter 3 ... 71

The greatest computer in the universe ... 74

Paradigm wars ... 78

Consciousness ... 83

What experiences the moment? ... 83

Is a top-down force theoretically possible? ... 85

Is there evidence for the top-down effect? ... 86

Exhibit #1—human knowledge. ... 86

Animal Behaviour ... 90

Mind over matter—the placebo effect ... 90

Mind over Matter—Dissociative identity disorder (DID) .. 94

Mind over matter—Brain Elasticity ... 96

Emergent properties pushing down ... 97

The agency of ideas ... 98

Does brain size really matter? ... 103

The conscious gatekeeper ... 105

Plugging Holes in Evolution and Deoxyribonucleic Acid (DNA) Explanations ... 105

Self-awareness ... 106

Limitations of the Mutations Strategy ... 107

- Down to earth about DNA 113
- Values 114
- Intelligent design 116
- Court cases about what is taught in US schools 118
  - Room for subspace 120
  - Not confined by the skull 121
  - Global Consciousness Project 123
  - Forensic evidence 125
- Psi 126
  - Probability 127
  - Effects 128
  - Telepathy 128
  - Remote viewing 132
  - Psychokinesis and the quantum measurement problem 135
  - Presentiment 140
  - The significance of Radin's experiments 142
  - Other results 143
    - Card predictions 143

Training research 143

Military research 143

Animals 144

Analysing psi evidence 145

Near-death experiences (NDEs) 148

Relationship between the brain and cognitive functions 151

Mind to mind, inter-dimensional communication 153

Witches, snake-oil salesmen and mediums 153

The Scole experiments 160

Other testing 163

Summary 165

Chapter 4 167

So what is religion doing at this science fair? 171

Meaning 173

From zero-sum to complementarity 176

Religion and subspace 178

Case studies 186

Discovering the science behind religion 193

Possibility of a single consciousness 199

Summary—triangulating religions 200

Chapter 5 203

The different parts 206

Triangulating the pieces 208

Ideas as emergent properties 212

Consciousness, the prime mover 213

A new paradigm 216

New models for our new paradigms 226

   Is it all just a hologram? 226

   A new integrated model—integral theory. 231

The role of the outsider 238

Appendix 243

Bibliography 247

Notes 253

Index 286

ACKNOWLEDGEMENTS

This is a special thank you to my wife Jacquie and boys, Kaemon and Kayode, for their beauty, wonder and inspiration. My foray into madness, AKA 'daddy's project', would have landed in a heap without them.

I owe all the scholars referenced in this book my gratitude for their unique, ground-breaking and often courageous work. A special thanks goes to the inspiring and courageous Dean Radin, PhD, Arthur Buehler, PhD, and Gerald H. Pollack, PhD. Hope comes easier with people like you around. A special thank you also goes to Barbara Wolff at the Albert Einstein Archives, Hebrew University of Jerusalem, for patiently putting up with all my nagging queries regarding what Albert Einstein actually said.

I also owe a great deal to those who have encouraged me and helped bail-out my boat when it was sinking. These people include (not in any particular order) my brother John, sister Natalie, my father Peter, the Zammits (especially Anne, Wendy, Victor, Lewis and Dorothy), Nick Tantaro, David Giezekamp, Andrew King, Rhonda Hunt, the Cheethams (Michael, Norm and Jock), Neroli Colvin, and of course Annette and Les Rogers.

Thanks to John Tiedemann for the great cartoons. To Angela Bell for her thoughtful editing and to Phillip Gessert for his wonderful design work.

And to my mother Stella and brothers Nigel and Adrian, thanks for your ongoing love and inspiration. If it wasn't for you this book would not have been written.

INTRODUCTION

Specialisation is a good thing. It refines our skills and explores a subject more deeply. As someone who has made a career out of managing complex projects I am aware of the benefits of specialisation, but also of its potential dangers.

Psychologist Abraham Maslow once said, '... it is tempting, if the only tool you have is a hammer, to treat everything as if it were a nail'.[1] We tend towards tools we are familiar with when confronted with a puzzle. Lawyers will lean towards legal solutions, computer technicians towards software and hardware solutions, psychologists towards psychological solutions etc. You can't blame them. That's how they've been trained.

Modern science has become highly specialised. One specialist area subdivides into others which in turn subdivide, creating a tree of specialisation.

Like our ancestral tree, each subdivision further removes us from our common origins. In our enthusiasm to keep the wheels of progress moving, the common origins can be forgotten or undervalued. Important connections become neglected.

Knowledge and research can become siloed and myopic. Areas which once worked together, like two sides of a coin, become competitors in a zero-sum game, each with its own increasingly technical language and systems. Specialist areas become competitors for finite resources, fuelling politics, which turn separations born out of convenience into philosophical divides.

Such a landscape challenges academics and other professionals in maintaining connections with other areas of science. Worse still, it makes it nearly impossible for the average person to feel any real connection with scientific exploration. Modern science has become too technical for the ordinary person. For many of us, understanding a Discovery Channel documentary is all we expect of ourselves when it comes to involvement in scientific discovery.

Modern technology has allowed us to delve so deeply into our specialisations that some areas are peeking through to the other side of existence, reaching the limits of what they can achieve on their own. This is producing intriguing behaviours indeed, such as physicists talking like spiritualists.

What if the universe we live in does not care to comply with the separations we have created? What if there exist phenomena so broad in nature that seeing and understanding them is beyond the capacity of one discipline alone but vital to us meeting our future challenges?

This book explores the possibility of *mega-phenomena*, phenomena that sprawl across our technical and philosophical divides but are concealed by our neglect of important connections.

There is no reason to restrict our thinking about the forms such mega-phenomena take. They could manifest a new law of physics or form of life. They could also be a solution to a long-standing problem, from global warming to conflict in the Middle East.

In the following pages we will examine one such mega-phenomenon which could be lurking behind some of our most profound questions. As far back as we can record, our intellect has been perceiving signs that what we can see, hear, touch and smell is not all that exists. We intuitively sense a timeless and spaceless dimension in parallel with our everyday world.

Like never before, technology is revealing tell-tale signs of this realm which begins to emerge when we bring together areas of knowledge that, for the last few hundred years, have become politically estranged.

Using the analogy of the fable *Six Blind Men and the Elephant*, we will take a *safari* through selected subject areas. We will inspect and review insights, research and discoveries that will help us *find the elephant*, the mega-phenomenon we will refer to as *Subspace*.

Through *triangulation*, a method of comparing different perspectives on a common subject, we will see how much of subspace can be revealed when we put together what modern scientific culture has pulled apart.

The process will also demonstrate how making connections through plain language can make science more accessible and reveal the value of different perspectives. It will show how aloofness in science is not only unnecessary but can be counterproductive.

This book will take you to some controversial areas in science; great breakthroughs and controversies are often bedfellows. It will not, however, explore the full breadth of the topic areas visited. While different views will be discussed, and references given for those who want to investigate them further, the book will generally identify complementary information and then move on. It will not engage in exhaustive debates on which school of thought is right about any given topic.

You will also notice that no matter how profound and innovative a place we arrive at, we quickly find Albert Einstein has already been there. A quote or example will be provided every so often, showing Einstein's 'trail of breadcrumbs' and how he is still way ahead of us, even though he is no longer with us. I have consulted with the Albert Einstein Archives, Hebrew University of Jerusalem[2], which hold copyright to Einstein's writings (in accordance with Albert Einstein's Last Will and Testament of 1950). With the help of the university, I will provide you with Einstein's most accurate quotes and/or translations, rather than the popular but dubious ones we often come across.

On a personal note, I am not affiliated with any scientific or religious/spiritual institution. In short, my career is not mixed up in this stuff. While I have my biases like any person, I have tried to use my perspective to provide an *outsider's* look into the subjects, to present alternate views to those generated by the politics between disciplines.

An outsider's perspective can be an asset to project management. Some of the principles of project management will be used in the discovery process[i]. This will not be a highly technical approach. It means following a process that builds our understanding of subspace by identifying complementary and confluent information. It is one which merely adds a little more discipline to how we live our lives each day, focusing on goals, adapting to new environments and being open to

---

i There are different forms of project management resembling this approach, such as *agile* or *iterative* project management (where the project moves forward as a discovery process).

the surprises life might have in store for us. It acknowledges that finding a solution means being open to the fact that at the time we embark, we don't know what the solution looks like. All we can hope for is that through a disciplined focus on our goal, we won't allow prejudice to blind us to the solution when it finally presents itself.

I am also aware that covering such diverse subject areas without a team of subject matter experts guiding my every move may mean errors on my part in the minds of some specialists. I apologise in advance for any such errors.

It is also worth noting that I often use the word *science* differently. Sometimes I mean institutionalised science ('the establishment', or what some people call *scientism*) but more often I mean the fruits of the *scientific method*. And one does not require a lab-coat to follow this. For me a scientist is someone who engages in the scientific method, which includes observation, predicting, testing and following the evidence. This could be anyone from a physics professor to a plumber at my kitchen sink.

The pages that follow are full of research and concepts from many different sources, not just academic ones. For example, although it does not have its own chapter, art is acknowledged through the book as playing an important part in human innovation. It comes in all forms, from sci-fi movies to Michelangelo's sculptures. You may find the references to popular art also help explain the scientific concepts being discussed.

For those topics you have some familiarity with, it will likely be an easy run. For areas outside your experience you might have difficulty following a topic every now and again. If you do discover a topic difficult to follow, don't get bogged down in it or give yourself a hard time. Just move on. There is much more you will understand and there is no one topic that is crucial. In fact, many of the ideas, including those related to quantum physics, surface again later in the book, in different contexts. You may understand them better then.

Of course this book is no *On the Origin of Species* ... or *Relativity, the Special and the General Theory*. Nor does it even claim to be a 'scientific' theory in any conventional sense. It is more a hypothesis, an invitation to consider new evidence and to look at some old and everyday evidence in a different way. It focuses more on the voices or examples of some of our greatest innovators, from Plato to Jesus to Newton to Wallace to Einstein, than the rhetoric of the scientific or religious institutions that today claim to represent their work. You may find that, as individuals, these people were more embracing of the breadth of the human experience than their descendent institutions (broad definition) might like to admit.

Unless you are already immersed in the topics to be covered, the information may challenge your current views about the world and reality. But this is not something you need to worry about.

Think of it as if you were coming to the end of your favourite TV series, only to find out you were in fact only halfway through it. Watching further episodes reveals the characters to be very different and more complex than you were led to believe. Some of the bad guys turn out to be good, and vice versa. You also find out that what you thought was happening was only the tip of an iceberg of even greater intrigue. How would you feel?

You'd feel great, right?

You may be fascinated by what is revealed when we challenge the politics of the knowledge industry. So sit back and enjoy the safari. Lots of balls are going to fall. Catch those you can. You are not required to catch them all.

CHAPTER 1
The elephant in the room

SPACE, THE FINAL FRONTIER

Space attracts our curiosity like few other phenomena. Space, and the universe that frames it, allows our imagination to roam free with possibilities, of what reality is or can be, and what the future might hold for us. We yearn to explore it, to find out if indeed reality can trump our imagination.

The problem is that the universe is a big place and to get anywhere requires travelling distances so great our brains have trouble comprehending them.

Some of our fastest craft reach fantastic speeds. The Voyager 1 probe, for instance, reached a speed of 62,120 km/h (just over 17 km/sec). That's roughly three-and-a-half times faster than the Space Shuttle travels in orbit.

As fast as you think 17 km/sec is, it is considerably slower than the speed of light, which is more than 17,000 times faster, roughly a billion (1,080 million) km/h. This works out to be 300,000 km/sec, what many physicists believe to be the interstellar speed limit, the fastest possible speed that can be achieved by anything.

Image of Voyager 1 launched in 1977. Source: PD-NASA. Effect added.

Yet even if we were to reach this speed, it wouldn't make us competitive space explorers. The closest planets likely to have life are well over 1,000 light years away;[1] that is, if we travelled at the speed of light, it would take us at least 1,000 years to reach them.

To merely cross our own galaxy at the speed of light, one galaxy out of billions, would take us 100,000 years. At the speed of the Voyager probe it would take us 1.7 billion years.

This pretty much makes any significant exploration of the universe impossible for us, at least based on our current high-speed technology.

Thankfully, this practical difficulty has not stopped science fiction writers harvesting the possibilities of the universe. Since the 1930s John Campbell, Isaac Asimov and other science fiction writers have got around the technical difficulties of long-distance space travel by proposing another dimension of existence that sits alongside our everyday dimensions but obeys different laws of physics.

Once a person or spacecraft gains access to this dimension, usually through some form of special technology, they are able to travel or send messages faster than the speed of light, sometimes instantaneously, across vast distances. In some cases it also allows travel through time.

The *Star Trek* series calls it *subspace* but it also goes by a number of other names, such as *hyperspace* in *Star Wars* and *Stargate*, *slipstream* in the television series *Andromeda*, or even the *time vortex* in *Doctor Who*.

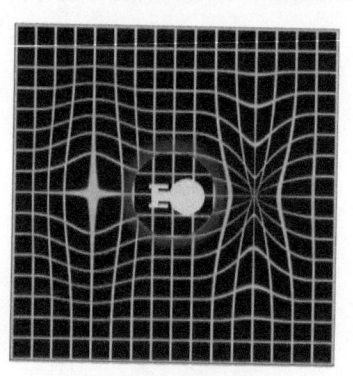

In Star Trek, a subspace corridor is a shortcut through space, which can allow a starship to traverse great distances otherwise unachievable with a traditional warp drive. A representation of a Star Trek 'Warp Bubble'. Source: Trekkyo623. Effect added.

With the help of imaginary machinery and highly technical mumbo-jumbo, people generally consider subspace plausible enough not to allow today's truth about space travel technology get in the way of a good space tale. After all, the existence of another dimension alongside our everyday one has been a strong suspicion of ours for thousands of years, from the religious afterlife concepts that inspired the building of the Egyptian pyramids to Plato's idea of a greater existence which most of us only see shadows of,[ii] where ideas have forms and represent the fundamentals of reality.

The existence of a dimension that would free us from the limits of time and space, and an ability to access it, would clearly open up many possibilities for the human race, if only it were true. If only science fiction could become reality.

The good news is that science fiction becomes reality all the time. Journeying to the moon, scuba diving, credit cards, robots, DVDs, CCTV, genetic engineering and the Internet all had their origins in science fiction. *Star Trek* technology, in particular, has been said to have inspired many of our modern devices, such as mobile phones, medical imaging, computers, SETI, MP3 players, Google Translator, global positioning system devices, iPods, virtual reality, focused ultrasound surgery and spaceship propulsion.[2] NASA recently discovered a planet which has two suns, first conceptualised by George Lucas in the original *Star Wars* movie, *A New Hope,* thirty-five years earlier.[3]

In Dr Who, the vortex is outside normal space-time, and therefore normal rules of physics do not apply. It is mostly used as a kind of 'wormhole' in time and space. In Dr Who it is accessed through the Tardis. Author: aussiegall. Effect added.

Could this be true for subspace as well?

Now more than ever, scientific evidence is suggesting that the existence of an accessible, parallel dimension with the properties of subspace may not be as far-fetched as some might think. In fact, we may

---

ii The ancient Greek philosopher Plato used the analogy of shadows on the wall of a cave caused by the light of a fire in *The Allegory of the Cave*, which can be found in Book VII of Plato's best-known work, *The Republic*—Plato (427–347 BC).

know more about it than we realise. Intriguingly, an ancient fable may hold the key to unravelling the mystery.

SIX BLIND MEN AND THE ELEPHANT

Six Blind Men and the Elephant.
Source: Illustration in the Golden Treasury Readers 1909.

Six Blind Men and the Elephant is an ancient fable which has surfaced in different cultures around the world.

In some versions a king (in others Buddha) calls six blind men together to describe something he brings before them.

Each grabs a different part, feels it, and then bickers with the others about the nature of the thing. Each assumes there is only one perspective (naturally, his own) and that others do not share the same perspective because they are inept.

Finally, the king interrupts the bickering to reveal that they were all a bit right. If they weren't so narrow-minded, they might have put their information together and realised that in fact they were each feeling different parts of the same animal: an elephant.

Is it possible there are phenomena like subspace, too broad to be captured by any one discipline? Like the blind men, each discipline captures a little of it, but it is only when different information is brought together, from separate disciplines, that *the elephant*, subspace, begins to emerge.

TRIANGULATION

This approach to uncovering phenomena that sprawls across our disciplinary divides is not new to science. *Triangulation* has been used for some time to gather knowledge about phenomena too broad to be captured by one discipline alone. It enables different types of knowledge to *join forces* to uncover phenomena bigger than any one discipline or knowledge-gathering method.

The first step in achieving the benefits of triangulation, however, is accepting that no one discipline or method of gathering knowledge is able to explain everything, and that different forms of knowledge can be *complementary*.

The technique has been particularly successful in combining quantitative (numerical—such as *how many?*) and qualitative (contextual—such as *does it feel comfortable?*) evidence to achieve clarity about such phenomena. Triangulation supports the view that all useful measurements require both qualitative and quantitative aspects, and that both dimensions should have equal status in describing reality.

Measuring the width of a river by triangulation. Source: Hulsius, 16th century.

There are other dualities which triangulation can help bring together, such as deductive and inductive reasoning. The former being the *top-down* logic; that is, reasoning from one or more general statements (premises) to reach a logically certain conclusion.[iii] This contrasts with inductive reasoning (*bottom-up* logic), where the conclusion is reached from specific examples.[4][iv]

Sociology is a field which has come to understand the need to consider both sides in any investigation into the true nature of things. Triangulation is used extensively in community development and the evaluation of social and development programs. In such complex situations, subjective and objective information is necessary to achieve

---

iii An example of deductive goes from general to particular, such as, 'All men are strong. Peter is a man, therefore he is strong.'

iv An example of inductive reasoning goes from particular to general. 'I met four men and each of them was strong; therefore, all men are strong.'

social and environmental outcomes. It acknowledges that human discernment, as much as statistics, is important to any solution that needs to be adopted and supported by people. We can see this principle at work in the criminal justice system, where statistics need to be complemented by human beings (judge as well as the jury) to get the right outcome.

> *'The water is 0 degrees'* (quantitative)
>
> *'The water is cold'* (qualitative)
>
> *'The light has a wavelength of 580 nanometres'* (quantitative)
>
> *'The light is yellow'* (qualitative)
>
> *'The man is shaking his hand in the air twice each second and his face is emitting light of a wavelength of 700 nanometres'* (quantitative)
>
> *'The man is red-faced angry'* (qualitative)

Triangulation allows different forms of research to work together in a way that takes advantage of the strengths of both.

While the technique has been used throughout history, it has recently been reborn in the triangulation metaphor, coming from its use in navigation to assess where on a map one is located.[v]

We use the principle of triangulation every time we search Google. By providing a number of words related to a particular subject, the search engine is able to collapse the billions of possible subjects on the Internet into a few that are most relevant, which contain all or most of the words you have provided.

Government and academic teams use triangulation to explore and improve their knowledge of the real world. It is a pluralist approach to knowledge, as it works off different views in order to cast light on a topic, cutting across knowledge types like the qualitative–quantitative divide.[5]

---

[v] One takes bearings of two landmarks in order to locate one's position. The angle between the two bearings, plus knowledge of the distance between the landmarks, allows the navigator to plot position.

Validation and complementary information

Triangulation has been used to validate information and reduce various forms of research bias in data. Apart from navigation, it is used in telescope arrays, which make use of many small telescopes spread over a large area to simulate the effect of one very large telescope. The array has the power of one very large telescope and has the ability to cross-check itself.[6]

Researchers Erzberger and Kelle talk about the other use of triangulation, in bringing together complementary information to discover an underlying truth:

'...the use of different methods to investigate a certain domain of social reality can be compared with the examination of a physical object from two different viewpoints or angles. Both viewpoints provide different pictures of the object that might not be useful to validate each other but that might field a fuller and more complete picture of the phenomenon concerned if brought together ...

Empirical research results obtained with different methods are like the pieces of a jigsaw puzzle that provide a fuller image of a certain object if put together in the correct way.'[7]

Triangulation proposes a common truth that is being perceived from different perspectives (for example, different people with different experiences). It is a bit like the blind men feeling different parts of the elephant.

Another technique employed in triangulation is *layering*. This is taking the research in stages, where at one stage you employ one technique, such as a qualitative method, at the next stage you employ another, for example a quantitative technique, one leveraging off the other.

In layering triangulation, one is not only using different perspectives in a common time reference, but using time to show different perspectives of that being investigated. This kind of makes sense anyway, because if something is true in an absolute sense (that is, true on a cosmic scale rather than our everyday scale), it should be as true yesterday as much as today. We may, however, have had a different view of it yesterday than we do today.

This serves as a reminder not to dismiss the perceptions of those who have lived and recorded observations long before us, just because they came before us. Indeed, the reverse may well be true; that is, our ancestors may have been able to see a side of something no longer available to us.

With triangulation comes the epistemological and ontological divide, the difference between what reality really *is* (ontological) and what we are able to say or *know* about it (epistemological). Acclaimed physicists Albert Einstein and Niels Bohr argued about this. Bohr would reply to Einstein's suggestions that physics is concerned with understanding nature, by saying that physicists can engage only in what they can say about nature. In this tradition JBS Haldane, a British geneticist and evolutionary biologist, in 1927 famously wrote that he suspected that the '...universe is not only queerer than we suppose, but queerer than we can suppose'.

On a universal scale, our brains are microscopic, so the idea that there are limits to what we can comprehend should not surprise us. But it is something else to suggest that there is no truth, as some philosophers, like some *deconstructionists*, might propose. No truth implies that everything is relative and nothing is real, leading to the concept that there is no knowledge, not even knowledge that there is no knowledge. It would also contradict evidence that there is a common truth that different perspectives can narrow in on, such as research related to group intelligence.

GROUP INTELLIGENCE RESEARCH

James Surowiecki writes about research into group intelligence and how crowds do better at decision making than individuals. In a 2004 article[8] he reminds us of the group work exercise where a number of people have to guess how many jelly beans are in a jar. It is based on a classic experiment in group intelligence in which, invariably, the group's estimate is superior to the vast majority of individual guesses.

When finance professor Jack Treynor ran the experiment in his class with a jar that held 850 beans, the group estimate was 871. Only one of the fifty-six people in the class made a better guess.[9]

This phenomenon is not just restricted to estimates of physical subjects. *Lost on the Moon*, a scenario developed by NASA, shows that a group of people are more likely than individuals (regardless of their profession or education) to arrive at the answers (confirmed by experts) on what to do if marooned on the moon with limited supplies.

Or consider the show *Who Wants to Be a Millionaire?* When a contestant is stumped by a question, he or she has a couple of choices in asking for help; that is, the audience, or someone he or she has designated as an expert. The experts do well. They get the answer right 65 per cent of the time. But the audience does better, getting the answer right 91 per cent of the time,[10] even though it's made up of 'regular Joes' in for a fun time at a TV studio, rather than technical specialists.

Experts do well, but the crowd does better. Host Vic Sotto asks contestant Joey de Leon in an episode of WHO WANTS TO BE A MILLIONAIRE? *Source: Howard the Duck.*

Similar effects have been shown in studies measuring the group's ability over individuals to pinpoint the weight of objects, the temperature in a room and even to predict future events (polling and horse racing).[11]

These examples appear to support the existence of a common reality which different perspectives can narrow in on. It suggests that triangulation-type techniques can provide information about all types of phenomena, from solid objects to future events.

Group intelligence research suggests different kinds of knowledge and experience can come together to shed light on a subject in a complementary way, whether quantitative or qualitative, deductive or inductive, mechanical or contextual etc. It shows that the different experiences, knowledge and approaches people have to knowledge can be a strength in knowledge-gathering rather than a weakness. In fact, the effect appears to be strengthened by the plurality of views; that is, the more perspectives, the better the outcome. The alternative proposition would be that there is no common reality and that the group actually *creates* reality, for example, the number of beans in the bean jar. Don't laugh too loudly. We shall return to this question later.

While there are principles researchers use to improve the results of triangulation, such as ensuring that there is a clear focus to questions and establishing the relevance of the methods to the questions being asked,[12] caution needs to be employed to prevent the technique becoming too prescriptive. Being particular about what knowledge type is relevant is open to prejudice. That is, by being too fussy about which knowledge types are allowed into the research, one could be making the same mistake triangulation was being employed to remedy.

THE INVISIBLE ELEPHANT

'The task is, not so much to see what no one has yet seen, but to think what nobody has yet thought, about that which everybody sees.'

Erwin Schrödinger[vi]

Some of the information which is going to be presented to you in the following pages may confront your assumptions about the world in which we live.

Valid and natural questions will arise for you when looking at this information: 'Why haven't we already been able to access and perceive this subspace dimension in our everyday life?' 'Why hasn't modern science already exploited these opportunities?'

Over the following pages you will encounter some possible explanations for this. But I think it is helpful at this point to remind ourselves that our understanding of the world is changing all the time. What we

---

vi Erwin Schrödinger, Austrian physicist who developed a number of fundamental results in the field of quantum theory, which formed the basis of wave mechanics.

thought about reality a thousand years ago is very different to what we think today. Indeed, what we think today is markedly different to what we thought 100 or 200 years ago. No matter what time a person lives in, there is a temptation for us to believe that our current time represents close to the final stages of scientific discovery.

In 1894 the revered scientist Albert A. Michelson,[vii] reflected a general feeling in the scientific community at the time when he said that 'the more important fundamental laws and facts of physical science have all been discovered, and these are now so firmly established that the possibility of their ever being supplanted in consequence of new discoveries is exceedingly remote...Our future discoveries must be looked for in the sixth place of decimals.'

Six years later Lord Kelvin,[viii] widely known for determining the correct value of absolute zero as approximately −273.15 Celsius, formulating the first and second laws of thermodynamics and being the first UK scientist to be elevated to the House of Lords, also represented a common view in the scientific establishment at the time when he said, 'there is nothing new to be discovered in physics now. All that remains is more and more precise measurement.'

Not long after these statements were made, two revolutions in science occurred

Lord Kelvin, Source: 1902 Popular Science Monthly Volume 61.

---

vii Albert Abraham Michelson was an American physicist known for his work on the measurement of the speed of light and especially for the Michelson–Morley experiment. In 1907 he received the Nobel Prize in Physics.

viii William Thomson (Lord Kelvin), 1900.

which turned classical nineteenth century science (the science the above two scientists were talking about) on its head: Einstein's General Theory of Relativity and quantum mechanics.

Today scientists suggest we know far less. Alex Filippenko, an American astrophysicist and professor of astronomy at the University of California, Berkeley, and member of the team that received the 2011 Nobel Prize for discovering dark energy (which is speeding up the expansion of the entire universe), suggests we are aware of less than four per cent of the universe.

'In the past couple of decades there has been a true revolution in our understanding of the universe. What we used to think of as being everything is actually only four per cent of the pie, if you think of a pie with slices. Four per cent consists of normal stuff of which we are made. And so you can see that our view of the universe as scientists changes...'[13]

While no one knows how much matter is in the universe, some modelling suggests that only 0.00000000000000000000042 per cent of the universe contains any matter[14]. So if this is the case, four per cent may be an overestimate.

All the more reason to keep an open mind about what might be out there and an equally closed mind about what isn't.

These historical misjudgements are not unique to the physical sciences and can be seen in all manner of human discourse. Consider the following quote:

'The children now love luxury; they have bad manners, contempt for authority; they show disrespect for elders and love chatter in place of exercise. Children are now tyrants, not the servants of their households. They no longer rise when elders enter the room. They contradict their parents, chatter before company, gobble up dainties at the table, cross their legs, and tyrannise their teachers.'[15]

Sound familiar? This is a quote attributed to Socrates by Plato about two-and-a-half thousand years ago. If things were that bad then, children should now be eating their parents. Have children really changed? Is it children who have changed or is it the adults? Or is it really a reflection of how we flatter our own judgements at whatever time we belong?

The old will always think the young get away with too much, scientists will always feel that it is close to the end of science, and doomsayers will always being suggesting that the end of the world is just around the corner.

Imagine for a moment you were 1,024 years of age. What would you say to the youth of today whom you suspected will also live to the same age? It is possible you might say the following?

Invest in real estate...

People have been saying 'the end of the world is nigh' for a long time now. Don't get too excited by it ...

And

In a thousand years from now, a lot of the things you thought you knew about the universe will be shown to be wrong, so keep an open mind ...

It does appear that the more we learn about our environment, the more we become aware of what we don't know. The latter shows no sign of diminishing, and in fact appears to be growing in proportion to our expanding awareness of the total picture. And then 'there are things we do not know we don't know', as Donald Rumsfeld said in 2002 while United States Secretary of Defence.

It seems natural for new ideas to be resisted. Pretty much every revolution in science, from the suggestion that the Earth is not the centre of the solar system to the relativity of time itself, had a hard time to begin with, with some facing fierce and vicious attacks from the scientific establishment at their debut.

Another quiz for you:

If you came across someone claiming evidence for an idea that challenged the fundamental way you saw the world, whose ideas were being labelled as 'non-science' and ridiculed by respected scientists, would you:

a. Dismiss the idea and join the scientific community in ridiculing the idea, suggesting that any evidence must be the result of either incompetence or fraud?

b. Accept the evidence because you had that suspicion anyway, and because scientists are often wrong about things?

c. Be intrigued by the idea and have a look at experimental results, the experiments themselves and whether they were repeated with similar results by others etc.

A and B answers are similar in that, while they are understandably the way people generally make decisions in their everyday busy lives, they don't follow the spirit of scientific investigation; that is, follow the evidence rather than preconceived ideas of what is being hailed as the *scientific* view at the time (in the following section you will review some research about how the brain works and why you can't trust such preconceived ideas).

For example, if you answered A, you would have been among those in history who ridiculed some of what we today consider to be science's most profound and celebrated discoveries. These include rocket-propelled spacecraft, warm-blooded dinosaurs, the law of conservation of energy, brains growing neurons, quantum physics, black holes, mosquitoes carrying malaria, dark matter, the Big Bang, the circulation of blood, the rotation of the Earth around the Sun (rather than the other way round), the curvature of the Earth, global warming, DNA, evolution, continental drift, light being simultaneously a particle and wave (more later[16]), just to mention a few.[17] Indeed, there are websites that list Nobel Prize winning discoveries that were at first ridiculed as *pseudo-science*[18] by 'scientific' opinion when they were made.

Even meteors had a hard time being accepted. It is for this reason that there is scarcely a single specimen of meteorites in European museums that predates 1790. The age of reason considered the phenomenon of rocks falling from the sky merely the rantings of ignorant superstitious peasants.

'The Enlightenment' period in the eighteenth century resulted in philosophers and scientists actively liberating themselves from superstition. Europe's leading rational authority at the time, Académie Française des Sciences, championed this cause. The idea that stones can fall out of the sky was denounced by the Académie as an unscientific superstitious delusion. Antoine Lavoisier, for example, the father of modern chemistry, announced, 'stones cannot fall from the sky, because there are no stones in the sky'.

Embarrassed museums all over Europe, wishing to be seen to be part of this enlightened Age of Reason, hurriedly threw out their cherished meteorite collections with the garbage as bad reminders of a superstitious past. Farmers who came to the Académie with samples of meteorites were laughed at and promptly shown to the door.[19]

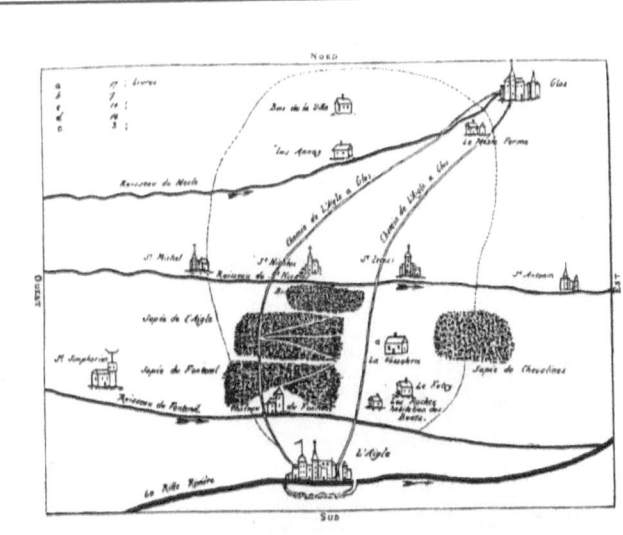

Académie Française sent the young scientist Jean-Baptiste Biot to investigate the spectacular fall of stones at L'Aigle in 1803. Biot, after an inspection on the ground, drew the map of the distribution of the fragments of the fall which was included in his famous report that finally allowed academia to accept meteorites as objects of extraterrestrial origin. We see the typical oval distribution of meteorites.

Towards the end of the eighteenth century some risked the ridicule to seriously research the subject, such as Peter Pallas and Ernst Florens Chladni. Things, however, didn't really change until the night of the 26th of April 1803, when the people of L'Aigle were awoken by the thunderous noise of thousands of rocks falling from the sky. The Académie Française was forced to take notice and appointed a commission that in the end reluctantly admitted that, indeed, it was possible for rocks to fall from the sky.

Across the Atlantic, they were not so easily convinced. In 1807, when two Connecticut scholars (one of them the chemist Benjamin Silliman) reported having witnessed a meteorite fall, President Thomas Jefferson (himself a scholar in the natural sciences) announced, 'I would sooner believe that two Yankee professors would lie than that stones would fall from heaven.' It wasn't until the 1833 Leonid meteor shower that American astronomers started to change their tune.[20] Fashionable prejudice about the inexistence of meteorites appears difficult to maintain when one's head and home start to get pounded by them.

Over the years the battles that these and other controversial discoveries had being accepted by mainstream science too often fall out of text books, in a way concealing the political, social and psychological barriers one faces when introducing new ways of looking at the world. I can understand how science may not want to promote examples of how unscientific it can be. However, glossing over this information does little to prepare the new generation for the challenges they will experience in the discovery process.

Understandably, intolerance to different ideas infuriates innovators. Einstein, for example, once commented:

'Great spirits have always encountered opposition from mediocre minds.[The latter are] incapable of understanding the man who refuses to bow blindly to conventional prejudices and chooses instead to express his opinions courageously and honestly.'[ix]

---

ix Einstein made this statement on behalf of Bertrand Russell, in March 1940, information provided by the Hebrew University of Jerusalem.

Born in 1823, Alfred Russel Wallace, who along with Charles Darwin developed the theory of evolution and natural selection, is known today as one of the greatest natural history explorers and thinkers of the nineteenth century. He was known for championing unconventional ideas which were often vigorously opposed by prominent scientists of his time (more on this later) which included his ideas about evolution and natural selection.

His other ideas included the role of geography in accounting for differences in species (he is known as the father of biogeography and zoogeography) and the concept of continental drift. He was also one of the first to raise awareness of the implications of the damage humans were doing to the natural environment, pioneer the first serious attempt by a biologist to evaluate the likelihood of life on other planets, launch research into mesmerism (now called hypnosis), and propose the merit of paper money. He was also known for raising awareness of what he considered the social ills of his times, including the arms race, European colonialism, the rise of the urban poor and the harshness of the criminal justice system, which he considered failed to reform criminals. He also pioneered the scientific study of spiritualism.

Alfred Russel Wallace 1878.

While today the merits of these studies are for all to see, in Wallace's time this was not so. In 1893, he reflected that his 'first lesson in the inquiry into these obscure fields of knowledge, [was] never to accept the disbelief of great men or their accusations of imposture or of imbecility, as of any weight when opposed to the repeated observation of facts by men, admittedly sane and honest. The whole of history of science shows us that whenever the educated and scientific men of any age have denied the facts of other investigators on a priori ground of absurdity or impossibility, the deniers have always been wrong.'[21]

Changing paradigms, as Einstein and Wallace have done—that is, changing strongly held foundational beliefs about the way the world is—is not easy to do. As we have seen, history is full of examples of how science has fallen short of its truth-seeking principles and its ideal image of an objective, evidence-loving community eager to challenge itself and embrace new ideas. Some have studied the dynamics of paradigm change.

American historian and philosopher of science Thomas Kuhn[22] famously described his view of the paradigm-change process:

1. People notice anomalies that the current paradigm doesn't explain.

2. People ignore the anomalies.

3. When the anomalies persist, people rearrange the current paradigm to try to account for the anomalies, making the paradigm increasingly clumsy and inelegant until …

4. A new paradigm is proposed, but is rejected.

5. Eventually a new paradigm is accepted and the preceding fuss is forgotten.

Kuhn had a more flattering view of scientists than others. Max Planck, one of the fathers of quantum physics, had a different view: '...a new scientific truth does not triumph by convincing its opponents and making them see the light, but rather because its opponents eventually die, and a new generation grows up that is familiar with it.'

I suspect that part of the problem relates to politics and power. When reputations and livelihoods have been built on a paradigm, politics will always be a barrier to change. As Upton Sinclair, a prolific American author and social change activist, put it, 'it is difficult to get a man to understand something, when his salary depends upon his not understanding it'.[23]

Kuhn used the duck–rabbit optical illusion to demonstrate the way in which a paradigm shift could cause one to see the same information in an entirely different way. Source: Jastrow, J. (1899). 'The mind's eye.' Popular Science Monthly, 54, 299–312

So the next time you are presented with the above quiz, consider answering C. It will not only make you more interesting at dinner parties, you will be more likely to find yourself on the right side of history.

IT'S NOT THEIR FAULT

'Common sense is the collection of prejudices acquired by age eighteen.'[24] [x]

In our quest for answers we will come across those unable to have their prevailing views challenged, despite the scientific evidence presented

---

x Mathematician Eric Temple Bell referred to Albert Einstein saying this in his book *Mathematics, Queen and Servant of the Sciences* (1952), p.42, although the Hebrew University of Jerusalem has no evidence of Einstein saying this.

to them. It will be natural for you to consider that this is because they are less intelligent, less educated, arrogant or narrow-minded.

Don't judge them too harshly though. Recent neuroscience suggests that this resistance to change may not be the fault of their mind, or their lack of intelligence etc., but due to the way their (and our) brain works.

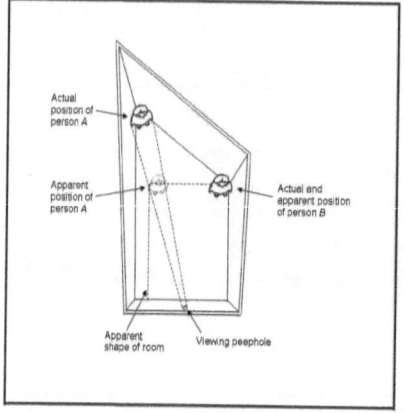

'Ames room' in Vilette science museum. Source: mosso—effects added.

In a recent TV documentary aired on the Australian ABC, *Redesign My Brain*,[25] Professor Bernd Lingelbach, displays his barn in southern Germany full of optical illusions, one of which is the Ames Room.

An Ames room is constructed so that from the front it appears to be an ordinary cube-shaped room, with a back wall and two side walls parallel to each other and perpendicular to the horizontally level floor and ceiling.

However, this is a trick of perspective, and the true shape of the room is trapezoidal; that is, the walls are slanted and the ceiling and floor are at an incline. The right corner is much closer to the front-positioned observer than the left corner (or vice versa).

As a result of the optical illusion, a person standing in one corner appears to the observer to be a giant, while a person standing in the other corner appears to be a dwarf. The illusion is so convincing that a person walking back and forth from the left corner to the right corner appears to grow or shrink.

The Ames room principle has been used widely in TV and movie productions for special effects, when it was necessary to show actors in giant size next to actors in small size. For example, the production of the *The Lord of the Rings* film trilogy used several Ames room sets in Shire sequences[26] to make the heights of the hobbits correct when standing next to the taller Gandalf.

Professor Bernd Lingelbach cites the room as an example of how the mind perceives through experience and has a tendency to construct reality according to what it feels it knows, or is familiar with. Our brains can be stuck in conventional *mindsets* that are difficult to get around. He shows that even after the illusion is explained, the mind continues to view the information the same way, even though it knows it to be deceptive.

It is as though, using Kuhn's example of the duck/rabbit, people will continue to see only a duck even when they are presented evidence that the picture is actually of a rabbit.

In the same program, Professor Allan Snyder, the director of the Centre for the Mind at the University of Sydney, shows how turning off the part of the brain that creates a mindset can improve the brain's ability to see through mindsets, be more creative and be better at solving puzzles.

Using electrodes, Snyder suppresses the left anterior temporal lobe, the area of the brain thought to be responsible for filtering and analysis as a result of past learning. He does this so that uncensored, unfiltered raw information can emerge from the right anterior temporal lobe, which is responsible for more creative and insightful thoughts and able to create new connections.

He provides demonstrations of how turning off the censor part of the brain (through electrical currents) enables the person to gain more insightful problem-solving skills.

Snyder explains, 'like any expert, you are blinded by your expertise...and that is the bottleneck to creativity...you are looking out at the world with the mindsets that you've accumulated. It's hard to break them.'

Jonah Lehrer writes in the *The New Yorker* about research suggesting that our level of intelligence doesn't save us from mindsets. Indeed, it shows that greater intelligence actually strengthens mindsets and makes it more difficult to see through them.[27]

Consider the following question:

A bat and ball cost a dollar and ten cents. The bat costs a dollar more than the ball.

How much does the ball cost?

Most respond quickly, insisting the ball costs ten cents. Of course this answer is wrong. The answer is five cents for the ball and a dollar and five cents for the bat.

Daniel Kahneman, a Nobel Laureate and professor of psychology at Princeton, along with others (for example, Amos Tversky and Shane Frederick), has been studying answers to these questions for decades. He shows that humans are not as rational as we like to believe.

When people face uncertain situations, they don't carefully evaluate the information or look up relevant statistics. Instead, their decisions depend on a long list of mental shortcuts, which often lead them to make foolish decisions. People tend to shortcut the process to avoid doing the necessary calculations. When asked about the bat and the ball, we skip the maths and instead reach for the answer that requires the least mental effort.

Although Kahneman is now widely recognised as one of the most influential psychologists of the twentieth century, his work was dismissed for years. He tells of one eminent American philosopher who, after hearing about his research, quickly turned away at a party, saying, 'I am not interested in the psychology of stupidity.'[28]

A recent study in the *Journal of Personality and Social Psychology*[29] led by Richard West at James Madison University and Keith Stanovich at the University of Toronto suggests that, in many instances, smarter people are *more* vulnerable to these thinking errors.

West and his colleagues began by giving 482 undergraduates a questionnaire containing a variety of classic bias problems, for example:

In a lake there is a patch of lily pads. Every day the patch doubles in size.

If it takes 48 days for the patch to cover the entire lake,

how long will it take for the patch to cover half of the lake?

The shortcut is to divide the final answer by half, giving you 24 days. The correct answer is of course 47 days.

West and his colleagues weren't only interested in reconfirming the known biases of the human mind. They wanted to understand how these biases related to human intelligence.

As a result, they interspersed their tests of bias with various cognitive measurements, including the SAT[xi] and the Need for Cognition Scale,[30] which measures the tendency for an individual to engage in and 'enjoy thinking'.

Surprisingly, the results showed that self-awareness was not particularly useful: 'People who were aware of their own biases were not better able to overcome them.'

This finding wouldn't surprise Kahneman, who admits in his book *Thinking, Fast and Slow* that his decades of ground-breaking research have failed to significantly improve his own mental performance: 'My intuitive thinking is just as prone to overconfidence, extreme predictions, and the planning fallacy.'[31]

Perhaps our most dangerous bias is that we naturally assume that everyone else is more susceptible to thinking errors, a tendency known as the 'bias blind spot'.

It originates in our ability to spot systematic mistakes in the decisions of others and our difficulty in spotting those same mistakes in ourselves. Although the bias blind spot itself isn't a new concept, West's latest paper shows that it applies across different types of biases. In each instance we readily forgive our own minds but look harshly upon the minds of other people.

He also found that intelligence seems to make things worse. The scientists gave the students four measures of 'cognitive sophistication'. As they report in the paper, all four showed positive correlations 'indicating that more cognitively sophisticated participants showed larger bias blind spots'.

---

xi The SAT is a standardised test widely used for college admissions in the United States.

This trend held for many of the specific biases, indicating that smarter people (at least as measured by SAT scores) and those more likely to engage in deliberation were slightly more vulnerable to common mental mistakes.

Education doesn't appear to be the answer, either. As Kahneman and Shane Frederick first noted many years ago, more than 50 per cent of students at Harvard, Princeton and MIT gave the incorrect answer to the bat-and-ball question.

A hypothesis put forward to explain these results suggests that the driving forces behind biases and the causes of our irrationality are largely unconscious, meaning they remain invisible to self-analysis and impermeable to intelligence.[32]

This understanding about how humans make decisions is moving into social policy. At an international conference I attended in June 2014,[xii] Harvard University Professors Cass Sunstein, David Laibson, Max Bazerman, Iris Bohnet and Mike Norton joined others including Professor Richard Thaler of the University of Chicago and Dr David Halpern of the UK's Behaviour Insights Team, to show how 'rational-choice theory', the foundation of much of modern economic, political and social modelling (which suggests people make rational decisions based on self-interest) is fundamentally flawed.

They showcased how governments were saving money and achieving greater social good by taking advantage of research revealing how people actually behave, which is more complex than rational-choice theory proposes. The approach, now used widely by the Obama Administration and UK policy units, coined as 'Nudge',[xiii] brings a scien-

---

[xii] *International Behavioural Insights Conference*, 2-3 June 2014, The United States Studies Centre, the University of Sydney, held at the Four Seasons Hotel, Sydney. http://bx2014.org/

[xiii] 'Nudge theory' was named and popularised by the 2008 book, '*Nudge: Improving Decisions About Health, Wealth, and Happiness*', by professors Richard H. Thaler and Cass R. Sunstein. The book is based on the Nobel prize-winning work of the Israeli-American psychologists Daniel Kahneman and Amos Tversky.

tific rigour and transparency to what marketing companies have been doing for many years; that is, observing how people actually behave and designing promotional activities in light of this. Though in this case for the purpose of positive social policy outcomes.

We all know examples of these little 'nudges' by business; for example, the nudges of fast-food outlets '...*would you like fries with that?*' These little nudges result in purchases of additional products (fries, for example), which translate into higher profits. Now governments are using similar mechanisms to achieve more helpful social outcomes, such as getting people to pay their tax on time simply by telling them that nine out of ten people do so. It has been able to increase the take up of pension plans by the lower-paid and the young, by adopting an 'opt out' rather than an 'opt in' process for enrolling in pension schemes. In other words, have people say they don't want it, rather than assume they don't if they fail to tick a box somewhere on a form. The approach has also shown how more unemployed people turn up to job club sessions by sending invitation letters that are more polite, positive and personalised. In one UK example, a sixfold increase was achieved in getting people to donate some of their estates to charities, simply by lawyers asking their clients if they want to do this when processing will documents (similar to the 'would you like fries with that' nudge).[33]

Nudge is turning the trick Would you like fries with that? into something good for you.

With the right oversight and transparency, this 'choice architecture', as it has been termed, provides, as Halpern suggests, a more humble approach to policy making. It makes room for us to say 'we don't know the answers'. We need instead to test, learn and adapt, and not rely on broad generalisations to determine how humans behave.

BEWARE FIRST PRINCIPLES

> *'Concepts which have proved useful for ordering things easily assume so great an authority over us, that we forget their terrestrial origin and accept them as unalterable facts. They then become labelled as "conceptual necessities", etc. The road of scientific progress is frequently blocked for long periods by such errors.'*
>
> Albert Einstein, Einstein's obituary for Ernst Mach, 1916.

One final note about the invisible elephant: it is important to remind ourselves that *'first principles'* in science have nothing holy about them. Even our so-called 'natural laws' are not really laws. As we will examine later, the universe has no rule book. Laws are made by humans, and the concept of natural laws is said to have originated from the religious view that God has laws over nature that can't be broken (more later).[34] In a real sense they are, as Einstein suggests, useful tools for ordering things easily, but we should guard against allowing them to accumulate undue authority over us or prevent us from seeing their limits.

In 1999 physicist Stephen Hawking gave a lecture on the possibility that space and time can warp.[35] He raises the possibility of utilising a subspace-like dimension for rapid journeys around the galaxy, or for travel through time. He provides theoretical arguments in favour of the existence of a subspace realm; but this is not the focus of the current example.

In his lecture he provided an example of how in the last century people began to realise that classical geometry 'laws'—like those of Euclidean geometry (that all the internal angles of a triangle should add up to 180 degrees)—are not always correct.

A triangle drawn across the surface of the Earth, for example, such as one which uses the tip of the North Pole and the line of the equator can have angles that add up to much more than 180 degrees. This is due to the curvature of the Earth; that is, the up-and-down third dimension.

It would have been unwise for science to reject the existence of triangles that measured internally more than 180 degrees because they did not fit into preconceived views of the world and geometry; for example, a formula created when the Earth was considered flat. Instead we must use the evidence to construct a different, more accurate view of the world as one with an additional dimension (displayed in the curvature of the Earth).

Hawking provided this as an example of how one cannot calculate the geometry of the world from first principles.

Instead, one has to measure the space we live in and discover its geometry by experiment. This

Albert Einstein warned of the dangers of holding on too tightly to the established rules or concepts of science. He showed how fallible they can be. Albert Einstein, 1931. Source: Library of Congress LC-USZC4-4940

means keeping a light hold on the so called 'laws' and being prepared to discover their limits. Indeed, as the eminent biologist Rupert Sheldrake suggests, our revered laws of nature, from the speed of light to the value of gravity, appear to change over time.[36] We shall discuss 'laws' in nature in more detail later.

As a case in point. A US scientist, Guiddo Fetta, invented a microwave space-flight thruster, the Cannae drive, which does not use propellant.

It is based on a similar device, the EmDrive, developed by the British scientist Robert Shawyer, who had a difficult time getting anyone to test it or take it seriously because it was thought to violate the laws of physics— in particular, the conservation of motion.

This did not stop China quietly testing its own version of it and apparently achieving positive results.

Fetta persuaded NASA to test his device and in 2014 its test team presented evidence that the thruster seemed to work, despite its apparent lack of respect for our laws of physics.[37]

NASA has attempted to explain the results:

'Test results indicate that the RF resonant cavity thruster design, which is unique as an electric propulsion device, is producing a force that is not attributable to any classical electromagnetic phenomenon and therefore is potentially demonstrating an interaction with the quantum vacuum virtual plasma.'[38]

Apparently, no one yet knows why it works.

As the Houston Space News Examiner blogger, Mark Whittington, suggests, the practical application of such a drive would present huge possibilities for space travel, 'larger spacecraft can be launched into space without the added weight of fuel and, because the thrust is low but constant, like an ion rocket, trip times throughout the solar system suddenly become weeks instead of months'.[39]

This is a good example of how, even today, we need to keep challenging the rules of science, even the holy-of-holies, in order to promote true innovation and discovery.

THIS 'IS SCIENCE' AND THIS IS 'NOT SCIENCE'

As history has shown, some of today's accepted and most celebrated discoveries in science were not considered *science* on their debut. Quantum physics suffered a similar experience during its first decades. Experiments by the physicist John F. Clauser on entanglement (now accepted widely in science) prompted the circulation of letters to educational institutions asking that Clauser not be hired, as he was involved in 'junk science'.[40] In the seventeenth century, Galileo's view that the Earth orbited the Sun, instead of the other way round, was also considered non-science.

And even now, the eminent biologist Rupert Sheldrake was invited to give a TEDx Whitechapel talk about challenging paradigms, *Visions for Transition: Challenging Existing Paradigms and Redefining Values*, in January 2013.[41] Sheldrake challenged the concept that today's science knows the principles of reality with only the minor details to be filled in (sound familiar?). He provided evidence that questioned some of science's most holy 'dogmas', including the assumption that the speed of light and laws of gravity never change and that evolution is a complete answer to life.

Sheldrake challenges the proposition that today's science knows the principles of reality with only the minor details to be filled in, Rupert Sheldrake 2008. Source: Zereshk, Effect added.

The anonymous TED advisory board took the talk off the TED website because, they claimed, it was 'not scientific'. It was only after an outcry from scientists and science enthusiasts all around the world that TED finally placed the videos back on one of its sites, but in a 'naughty corner', as Sheldrake calls it. A talk by Graham Hancock on consciousness at the same event was given the same treatment.

You can find Sheldrake's response to the seven reasons provided by TED for his special treatment. Hear an interview on YouTube.[42]

Some claimed that the TED board was heavily influenced by evangelical atheists, such as PZ Myers, who is known for his inflammatory writings and stunts, such as ripping out pages of the Qur'an, piercing them with a nail, throwing them in the trash with coffee grounds and a banana peel, and photographing the scene for his blog.[43]

Ironically, TED's treatment of Sheldrake's talk provided many with everything but confirmation that dogma (in the form of politics) had too tight a hold on today's scientific establishment.[44]

It illuminates the fact that what is considered 'science' can be as much a political as a technical assessment. It reminds us of other words that have become highly politicised, such as 'democratic'. Just because, for example, some people call something democratic, such as 'the Democratic People's Republic of Korea' (better known as 'North Korea'), doesn't mean it is. In the same way, people can call something *scientific* (for example, Japan's whaling program), or indeed *religious* (for example, Scientology or the Islamic State), but it doesn't make it so. In fact, one needs to be doubly suspicious when an institution feels the need to wrap itself up in words dripping with such politics.

Like *democracy*, the term *science* is often used as a value judgement, to legitimise 'acceptable knowledge'. As people like Galileo will tell you, this is knowledge that maintains a way of looking at the world that supports the current political power structures at a particular moment in history. It is these powers that maintain stability, peace and social order.

Peace and order are important too. Civilisation was hard won. However, we all have a role in keeping society moving along the road of truth as delicately and peacefully as possible, understanding that politics—that is, people knowing 'which side their bread is buttered on'—will tend to affect opinion about what is, and what is not, *science*.

SUMMARY

Subspace, a realm in which time and space do not present the same kinds of barriers as they do in our everyday world, has been conceptualised in our imaginations. To access and navigate such a dimension would propel the human race into an age of immense possibilities.

There is reason to believe that subspace may exist, and that it is only science's tendency to over-specialise that hides it from our view. It may be an example of a phenomenon too broad to be captured by one discipline alone.

Triangulation is a method familiar to science which brings different types of knowledge together to shed light on phenomena that sprawl across the disciplines we have defined. But in order to realise the potential of this approach we need be realistic about the huge challenge involved in bringing different fields of knowledge together, and changing the way we might think of the world. The fable of the *Six Blind Men and the Elephant* reminds us of what will happen if we don't control our prejudices.

Beyond prejudice we will also need to understand that many intelligent and opened-minded people will still not see the 'forest for the trees', as our brains can be a 'two-edged sword'; it can inform as well as deceive even the best of us. And then there is politics, knowing 'which side your bread is buttered on', which, as history has shown, can affect our individual and collective perceptions like few other phenomena.

Things are changing all the time. It is better not to hold on too tightly to those fundamental 'laws' or tenets. For example, all fish are cold-blooded, right? Wrong. Scientists have just found one that isn't (the opal).[45] There goes another 'rule'.

To pave the way for real progress we may have to get more comfortable with the idea that we are some way from truly understanding the existence we find ourselves in. This may be more difficult for some than others. There is a natural need to feel some control over our world, and our ability to accurately comprehend it is an important part of this.

Many of us feel just smart enough to know how 'dumb' we are, yet even this is likely to be a gross overestimate of our intelligence.

Collectively, we also may at some level prefer not to know how 'dumb' we really are. We may deny it to ourselves. But perhaps we can replace some of this insecurity with a pride in what we have learnt, an adventurer's appreciation of the mysteries yet to be solved, and a little more patience with ourselves.

DISSECTING THE ELEPHANT

Before we put the *elephant* together, we need to understand its parts. The *blind men,* as described in the fable, need to grab a part and use their senses to extract as much information about it as possible.

For simplicity and to stick with the triangulation metaphor, let's break the elephant into three parts: the head, the torso and the legs.

Let us pretend that the legs of the metaphoric elephant represent the study of physics; the torso, biology; and the head, religion or spirituality.

CHAPTER 2
Physics—the legs

PHYSICS—THE LEGS

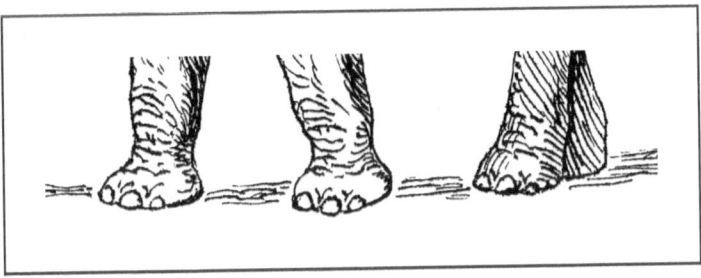

Let us consider physics as simply 'the scientific study of matter, energy, space and time and the relations between them'.[1] Does a review of physics research tell us anything about a possible dimension beyond our everyday world, a dimension which has different relationships with time, space and matter?

NON-MATERIAL REALITY—THE NATURAL FORCES

One of the first things physics shows us is that not everything that exists can be seen or touched, and that the material world we are familiar with (dominated by matter[xiv]) is only one aspect of our reality.

Though no one knows where its energy comes from, gravity is one of the primal forces that built the universe. Yet it can't be seen or touched. It is an immaterial enigma. We still don't know what gravity is.[2]

---

xiv The term *matter* is being used here to refer to corporeal substances; that is, tangible things which can be large or incredibly small but fundamentally *solid* (including gas and liquid molecules), which in principle could be felt by the touch or smelt or hit with another object, that exists in one spot at a time and has some mass or weight. There are other definitions of *matter*, which include phenomena that originate from what we refer to as corporeal matter, though this broader definition is not being used in this book.

Magnetic fields and the nuclear forces also affect us and our environment, yet they too are immaterial.[3] Electromagnetic waves, such as X-rays and radio waves are an everyday part of our world, and they too are immaterial.

We experience the signs of these forces and phenomena every day, such as the force of gravity holding our bodies to the Earth, and the electromagnetic signals which make our radios, mobile phones and Internet work, or which give us the experience of colours and warmth.

The fact that they are immaterial doesn't make them less real. We accept they exist without question. A non-material reality is therefore not at odds with our experience of reality, but a fundamental part of it.

So much so that a renowned Harvard University professor of physics, Lisa Randall, thinks that the relationship between the four fundamental forces of nature provides us with a clue as to the existence of an unseen dimension.

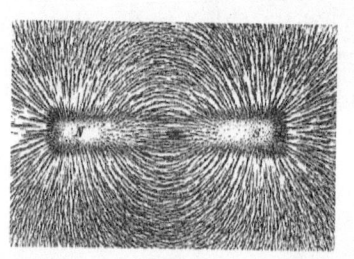

The immaterial magnetic field of a bar magnet revealed by iron filings on paper. A sheet of paper is laid on top of a bar magnet and iron filings are sprinkled on it. The needle-shaped filings align with their long axis parallel to the magnetic field. They clump together in long strings, showing the direction of the magnetic field lines at each point.

All forces in nature may be classified into four types. The *strong nuclear force* holds an atom's nucleus together—the protons and the neutrons—and is the power released in a nuclear bomb.

The *weak nuclear force* is responsible for certain types of nuclear reactions; for example, it triggers radioactive decay.

The *electromagnetic force* governs atomic-level phenomena, binding electrons to atoms, and atoms to one another to form molecules. For example, this force can be exhibited in magnets.

Finally, *gravity* is the force which holds together the universe at large, including the atmosphere, water, and us to planet Earth.

The curious thing is that the first three forces have roughly the same strength while gravity is much weaker. In fact it is a trillion, trillion, trillion times weaker.

For Randall, the observation that a small magnet can pick up a metal paper clip (electromagnetism), overpowering the force of the whole Earth (gravity), provides a tell-tale clue.

With a paper clip in her fingers, Randall states, 'if you think about it, the force of magnetism that is exerted on this paper clip is enough to compete and overwhelm the force of gravity that is acting on it. So there is a mystery there, because why is electromagnetism so much stronger than the force of gravity?'[4]

To Randall and her colleagues, the exceptional weakness of gravity compared with the other forces is a possible sign of another dimension beyond our immediate senses, where gravity itself is more concentrated.

Using *String Theory* (more on this later), Randall's mathematics suggest that there may be another existence in parallel to our own, where gravity is much stronger, compressing space trillions of times smaller, warping the dimension connecting our two existences.

Randall suggests gravitons (hypothetical elementary particles that mediate the force of gravitation) flow to our existence from this other realm. However, as we have an expanded space dimension, the gravitons are less concentrated, as they are spread more thinly through space. This would, says Randall, account for the weakness of gravity relative to the other forces.

QUANTUM MECHANICS

As Stephen Hawking and other prominent physicists will tell you, quantum mechanics is the most tested area in science, and has never failed a test.[5] Along with Einstein's General Theory of Relativity, it is considered one of the great revolutions in science and has shed light like never before on the basic properties of reality.

This field has provided insights into the existence of a subspace-like dimension through numerous experiments. Firstly, it has shed light on the true nature of matter.

Our everyday world has a very strong reliance on matter; that is, objects that one can feel, touch and smell. But findings in quantum physics suggest that matter isn't what it appears to be.

NATURE OF MATTER

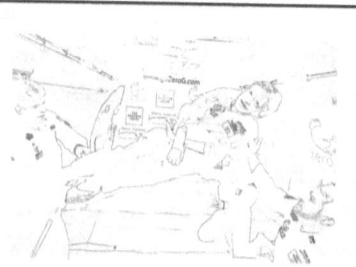

Because our atoms are 99% empty space, we are really quite hollow, even ghostly, in essence. We are hardly here. Noted physicist Stephen Hawking (centre) enjoys zero gravity during a flight aboard a modified Boeing 727 aircraft owned by Zero Gravity Corp. (Zero G). Source: NASA Jim Campbell/Aero-News Network. Effect added.

As quantum physics delves deeper into the subatomic world, humans have gradually come to understand that atoms are 99 per cent empty space.[xv] What is not empty space is made up of smaller particles (like *quarks*) and wave forms. These are made up of even more bizarre vibrating phenomena that pop in and out of existence and can't be measured or pinned down to any specific location.[6] The more you drill down, the less there appears to be.[7] As Niels Bohr, a Danish physicist who made significant contributions to understanding atomic structure and quantum theory, once said, 'everything we call real is made of things that cannot be regarded as real'.

Indeed, the vast majority of 'matter' in the universe is termed 'dark matter', which we know virtually nothing about (even though more of it is appearing). 'We don't know what it is made up of and where it comes from,' says Brian Schmidt, Australia's 2011 Nobel Prize in Physics winner.[8] Most of the universe, says Schmidt, is made up of dark

---

xv As the early twentieth century British physicist Sir Arthur Eddington put it, 'Matter is mostly ghostly empty space. To be more precise, it is 99.9999999% empty space.' (Russell 2004).

energy – 'dark' because we know as little about it as we do about dark matter[9].

Tim Tait, a physicist at the University of California, Irvine, says that 'in the recent years, we have become really aware of the fact that when we account for all the stuff in the universe, there is stuff that is missing. You can see it pulling on other things gravitationally, but other than that it doesn't leave any trace that it is there.'[10]

Tait belongs to a group of physicists who think that dark matter may actually be matter which has moved into an additional (fourth) dimension. Through high levels of energy, perhaps coming from early moments in the Big Bang, matter might have been pushed over the threshold into this other dimension. We can see its sign in the universe but can't perceive it the same way we do everyday matter within our three-dimensional view.[11]

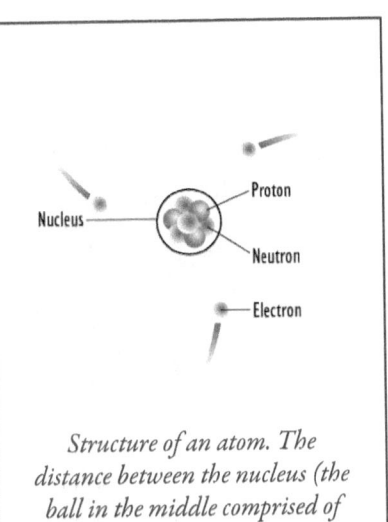

Structure of an atom. The distance between the nucleus (the ball in the middle comprised of neutrons and protons) and the orbits of the electrons is relatively great, making an atom mostly empty space. To the scale of this picture, the electrons could easily be orbiting a suburb away from you. The number of protons, neutrons and electrons determine what element it is, for example, gold or carbon. Gold has more of them, which is why it is heavier than carbon.
Source: User:Fastfission.

Researchers are trying to find out what exactly these little things are that make up all matter, including dark matter, but generally agree that they are not tiny versions of what we consider 'matter' or material objects. As the German physicist Hans-Peter Dürr put it, 'matter is not made of matter'.

MATTER IS AN EMERGENT PROPERTY

Many would be aware of the research being conducted around 'emergent' properties in living systems,[12] where the whole has different properties than the sum of its parts. This phenomenon is being observed in different situations, from the behaviour of the mind (for example, Gestalt psychology) to group human behaviour, to 'super-organisms' like bees and ant colonies.[13] The mechanics of how these emergent qualities come about is still a mystery. Our current senses can't detect the agency at play when emergent properties are engaged, even though we can see its signs, such as the amazing group behaviour of super-organisms like ants, abilities that can't be simply explained from a genetic programming point of view.

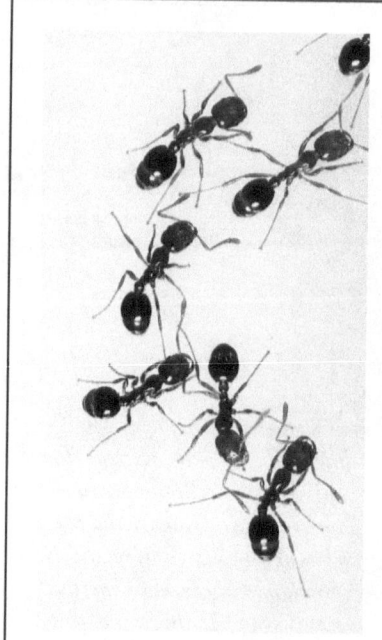

Super-organisms behave as a single organism when they reach a certain number. A new entity emerges. Ants and bees are typical super-organisms. The image is of fire ants, Stephen Ausmus, effect added.

Quantum physics is also telling us that this phenomenon describes the properties of matter; that is, far from being an irreducible ground-zero component of the universe, matter is also an emergent property of something else.

As Hawking et al. explains, Newton's theories describe the way in which particles (made up of the above bizarre phenomena) behave when brought together in large numbers to make composite structures like rocks, jumbo jets and ... us. 'What we do know is that components of all objects obey the laws of quantum physics, and the Newtonian laws are a good approximation for describing the way macroscopic objects made of those quantum components behave' (Hawking and Mlodinow[14]).

Even more bizarre are the mathematics coming out of black hole physics. It suggests that everything we experience is much more like a hologram than a picture of reality. 'Is the three-dimensional world an illusion?' asks the Professor of Theoretical Physics at Stanford University, Leonard Susskind, '... I'm inclined to think yes'.[15] The mathematics, says Susskind, hints at a two-dimensional reality at the surface of the universe, projecting the illusion of a three-dimensional reality which we experience.

Research into the components of reality suggests a 'real' world quite different to the one we experience, whose nature is essentially a manifestation of energy, an effect created out of billions of subatomic phenomenon 'in formation'. The specifics of this formation, its arrangement and design, provide us with a material world to explore.

All matter is made up of the same basic phenomena. It is the arrangement of phenomena which makes different materials; for example, the way quark particles form neutrons and protons and how they combine with electrons to make different elements.

Change the design and you can change iron into gold. This is what stars do. Not movie stars. But stars in space. Their furnaces transform one element into another by altering the number of electrons, protons and neutrons. The gold in your necklace and the carbon in your body were all made in a star somewhere out of, originally, hydrogen atoms.

Given that the material world we are familiar with is a representation of a different reality at a fundamental level which pops in and out of existence, the possibility of a reality beyond what we can see, touch and smell is not only plausible, but arguably a much closer approximation to true reality.

This information suggests that far from being a subordinate to our reality, what we call subspace could be the more primal plane of existence. One could even say that there is nothing 'sub' about subspace. Our everyday world may, in the fullness of time, turn out to be the real '*sub*-space'.

ENTANGLEMENT

Entanglement refers to the phenomenon observed in small objects which, after having interacted with each other, become *entangled*, and act as if they are physically connected, even though they are separated by distance. In this way they are said to exhibit 'non-local' behaviour, which defies the separation we usually experience between space (distance) and time.

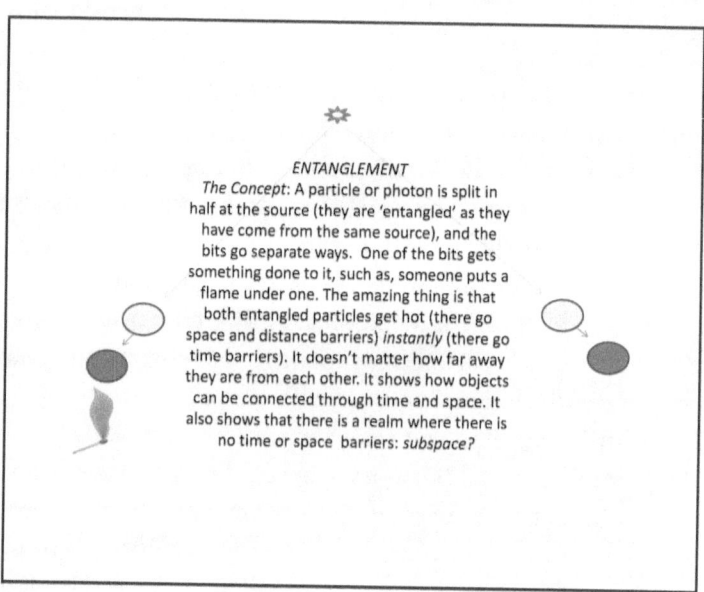

ENTANGLEMENT
The Concept: A particle or photon is split in half at the source (they are 'entangled' as they have come from the same source), and the bits go separate ways. One of the bits gets something done to it, such as, someone puts a flame under one. The amazing thing is that both entangled particles get hot (there go space and distance barriers) *instantly* (there go time barriers). It doesn't matter how far away they are from each other. It shows how objects can be connected through time and space. It also shows that there is a realm where there is no time or space barriers: *subspace?*

In essence, if you do something to one of them, they both respond in the same way. It doesn't matter whether the objects are a metre apart or 100 kilometres apart, both objects react to the stimulus *instantaneously*,[16] or at least at 100,000 times the speed of light, as some research has suggested.[17]

Why don't we see this in our everyday world? Because the effect is better observed on small objects, due to the interference of many entanglements experienced by larger objects. But theoretically it can occur in larger objects as well.

Research into quantum entanglement was initiated in 1935 in a paper by Albert Einstein, Boris Podolsky, and Nathan Rosen, and then by Erwin Schrödinger. Einstein called it 'spooky' action at a distance. Although these first studies focused on the counterintuitive properties of entanglement, with the aim of criticising quantum mechanics, eventually entanglement was verified experimentally.

No other forms of communication between the two objects exist to account for the effect. Bell's Theorem (acclaimed as the most profound discovery in all of science[18]) has shown mathematically and experimentally that, contrary to Einstein's suspicion, the effect is not the result of an undiscovered variable—the non-local connection is real.

In 2007 the entanglement effect was shown over a 144 km distance,[19] although distance appears to make no difference to the results. The phenomenon has been so well tested that researchers are no longer concerned with verifying it but investigating its use for communication, computation and even teleportation.[20]

This evidence is important to our quest for subspace. It shows that within one dimension of an object's existence, it can maintain a connection with another object through distance or space. For some reason this dimension is masked from our everyday world, and our normal senses. This tells us that the subspace described in science fiction, where distance is not the obstacle it is in our ordinary world, does appear to be supported by modern science.

> **EINSTEIN ATTACKS QUANTUM THEORY**
>
> Scientist and Two Colleagues Find It Is Not 'Complete' Even Though 'Correct.'
>
> **SEE FULLER ONE POSSIBLE**
>
> Believe a Whole Description of 'the Physical Reality' Can Be Provided Eventually.
>
> *Newspaper headline from 1935, regarding Albert Einstein and the release of his paper about entanglement with Podolsky and others.*

SPACE-TIME

While entanglement might seem strange, it makes more sense when you consider the nature of space and time as essentially manufactured products. As Brian Schmidt explains, space is being created all the time. It accounts for the expansion of the universe and explains why the universe is expanding faster than the speed of light.

It is not because interstellar objects are going faster than light. It is because space is being created in between them, says Schmidt. So much that objects in space are moving apart at a relative speed greater than that of light. Schmidt suggests that in the future the universe will become a big, but dark, place. We will lose sight of the other galaxies as more and more space is created between us and them.[21]

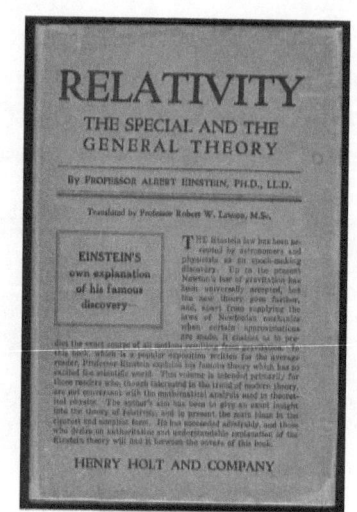

Einstein proposed his theory of relativity in 1905. There were two main parts of his theory. First, Einstein discovered that the speed of light is constant. Second, Einstein determined that space and time are not absolutes; rather, they are relative to the position of the observer.

The original 1920 English Publication of the paper.

Past, present and future continuum

Einstein's Special Theory of Relativity has been tested thoroughly since it was validated in 1971 by an experiment involving a jet and two atomic clocks. It supported Einstein's theory that we all experience time differently and that our relative speed and the strength of gravity affect the passing and experience of time. In fact, the relationship between time and space is a little like north and west, says Brian Greene, a theoretical physicist at Columbia University—'the more you get of one, the less you get of the other'.[22]

'For us believing physicists, the demarcation between past, present and future has merely the significance of but a persistent illusion.' Albert Einstein. Pictured during a lecture in Vienna in 1921—Source: Author Ferdinand Schmutzer.

The faster I move, the more I am travelling in time differently to someone stationary. If I go towards you I will be moving into your past, and if I move away from you, I move into your future, suggests Greene.[23]

Because of our proximity to each other we do not notice our relative time travel. However, if we lived far away from each other—for example, on different sides of the universe—my motion would be teleporting me hundreds of years into your future or your past, depending on which direction I was moving.

Steven Spielberg brought time travel into our living rooms with the successful Back to the Future movie trilogy. He brilliantly described the complexities of time travel in a way we could all understand. The image is of a replica Delorean DMC-12 Time Machine, the machines used in the movies to travel through time. Source: Ed g2s. Cropped and effect added.

One of the implications is that the future and the past exist in a real sense, as much as the present does. As Einstein put it to the family of his friend Michele Besso upon Besso's death, 'He has departed from this world a little ahead of me. That means nothing. For us believing physicists, the demarcation between past, present and future has merely the significance of but a persistent illusion.'[xvi] [24] As Max Tegmark, cosmologist and physicist at MIT, suggests in Brian Greene's, *The Fabric Of the Universe*, [25] the 'past is not gone and the future isn't non-existent ... past, present and future are all existing in the same way'.

If you think about it, it kind of makes sense anyway; it would be difficult to imagine that all of the universe is wiped out of existence each fraction of a second after our present, and then quickly reproduced again (in strikingly exact detail) a fraction of a second before we perceive it (to provide that continuity of experience).

Why now?

There is 'nothing in the laws of physics that picks out one now over another now' says Tegmark.[26] There is no necessity for us to experience only a portion of time at once (the present), and there is no require-

---

[xvi] As correctly translated by the Albert Einstein Archives, Hebrew University of Jerusalem. Note the difference between this translation and the translation often incorrectly used, that is, instead of 'believing physicists', it is inserted 'People like us, who believe in physics ...' Some may have trouble with the thought that Einstein was a 'believer'.

ment for us to perceive the passing of time in the direction that we do (the arrow of time pointing from the Big Bang into the future).[27]

Modern physics also says there is no reason why time flows in the direction it does. The dynamics of the universe can work just as well in reverse; that is, a causation going from the future into the past. As Peter Galison, professor in the history of science and physics at Harvard University, suggests, the mathematics of physics confirms 'there is a contradiction between the physics which seems fundamentally reversible and so much of our life that seems irreversible'.[28]

(Later we will consider the phenomenon of consciousness and its possible role in explaining our experience of a particular moment in time.)

This awareness of the relationship between time and space is not new. Einstein realised in 1905 that space and time were intimately connected with each other. One can describe the location of an event by four numbers. Three numbers describe the position of the event together with a time location (up-down, back-forward, right-left and before-now-later), that is, *space-time*. Time is the fourth dimension, as stated by Einstein himself.[29] Many physicists believe that space-time, like the natural forces, is fallout from the Big Bang, and that it has been in production ever since.

The existence of reality across time, our ability to travel in time (our everyday experience of time), our ability to regulate the speed of our experience of time and, indeed, speed up or slow down our individual travel in time (though be it across vast distances in any practical sense) shows that space and time are malleable. They are not absolutes; that is, pillars of reality itself. This points again to a more primal subspace from which time and space emerge.

As objects, we too would appear to have an existence in this subspace realm. The question is, how does this manifest in our experiences? How can we sense this other dimension?

FINDING THE ELEPHANT IN STRING THEORY AND WORMHOLES.

String Theory suggests that elementary particles, like protons and electrons, could be thought of as 'musical notes' or elementary strings stretched across dimensions. It is a contender for the Theory of Everything (TOE), a mathematical model that describes all fundamental forces and forms of matter. In the early 1990s there were five different string theories, each suggesting there are *ten dimensions*, not just the four (including time) we experience.

Six of these ten dimensions are said to be curled up into a space so small that we don't notice them. The remaining four dimensions are fairly flat, and are what we call space-time, the dimensions of our everyday world. However, all five theories appeared to be in competition and contradictory.

Edward Witten's brilliant mathematics created M-theory, showing how thinking more deeply can solve what seem to be contradictory propositions in sting theory. He 'FOUND THE ELEPHANT' *in string theory.*

In 1995 at the University of Southern California Conference for String Theorists history was made by Edward Witten, a mathematician at the Institute of Advanced Study, acclaimed as perhaps the only living successor to Einstein, due to his superior understanding of mathematics.

Witten's presentation showed that by expanding our thinking and considering the possibility of eleven dimensions, all the five conflicting string theories could be seen as only different aspects of the same phenomenon. He had *found the elephant*. Its effect on theoretical physicists was like the Sermon on the Mount on Christians. *M-Theory* was born, and today it is still considered by many, including Stephen Hawking, as the closest we have come to a Theory of Everything.

Wormholes, time warps and warp drives

Four years after Witten's presentation, in 1999, Stephen Hawking gave a lecture on space and time warps.[30] As discussed, science fiction programs such as *Star Trek* use time warps or warp-drive engines to access subspace and to journey around the galaxy or through time.

Touring through Einstein's General Theory of Relativity, Kurt Gödel's mathematical modelling and more recent work around cosmic string theory, Hawking explains how we have experimental evidence of the bending of light and the curving of space-time.

Artist's depiction of a hypothetical Wormhole Induction Propelled Spacecraft, based loosely on the 1994 'warp drive' paper of Miguel Alcubierre. Source: NASA CD-98-76634 by Les Bossinas.

To create a wormhole, says Hawking, one needs to warp space-time in the opposite way to how normal matter warps it. Ordinary matter curves space-time back on itself, like the surface of the Earth. However, to create a wormhole requires something which can warp space-time in the opposite way, perhaps with the use of negative mass, and negative energy density.

Hawking explains that Quantum Theory has overthrown classical laws to allow energy to be negative in some areas, provided it is positive in others, through the help of the Uncertainty Principle (the more you know about the speed of a particle, the less you know about its location).[31] This requires fields, like electromagnetic fields (radio, light etc.), to have a certain amount of fluctuation and not be allowed to fall to zero, not even in the vacuum of space.

In that vacuum, pairs of particles and anti-particles suddenly appear together, move apart, and then come back together again, annihilating each other in the process.

The *Casimir effect* shows how negative energy can be created between two parallel metal plates a short distance apart. The plates allow only particles at specific frequencies between the plates. This means that particles that would normally occupy that space cannot do so, resulting in a dearth of them, relative to everywhere else besides the space between the plates.

The dearth causes a negative vacuum between the plates, which creates a slight force pushing the plates together. This has been measured experimentally. As the energy density far away from the plates must be zero in space, the energy density between the plates must be negative. The experimental evidence that space-time is curved together with the Casimir effect confirms for Hawking that we can warp space-time in the negative direction. It lays down the possibility for future space-time leaps and travel by bending dimensions.

Of course, there would be a lot of practical problems to consider if this were possible and Hawking delights at providing his views on these. But for the purpose of this subject, it is enough to show that the existence of a dimension outside of our own everyday ones, including the possibility of accessing it to achieve the advantages of subspace, is far from fantasy, but serious science and well within the realms of possibility.

Since then NASA has started research into developing a warp drive. In September 2012, scientists, engineers and space enthusiasts gathered at the Hyatt Hotel in Houston, Texas, for the second public meeting of 100 Year Starship, run by a former astronaut, Mae Jemison. Its aim is to 'make the capability of human travel beyond our solar system to another star a reality within the next 100 years'.

Harold 'Sonny' White of NASA, who runs the advanced propulsion program at Johnson Space Centre, gave a presentation on 'Warp Field Mechanics 102'. White specialises in warp-drive research for NASA. He outlined the physics of a potential warp drive, describing how he has computed theoretical results that could pave the way for an actual warp drive and announced that he was commencing physical tests in his NASA lab, which he calls Eagleworks.

White says, 'a trip that with current technology would take 75,000 years, a ship with warp drive could make in two weeks ... perhaps a *Star Trek* experience within our lifetime is not such a remote possibility.'[32]

LIGHT AND THE ROLE OF THE OBSERVER

Thomas Young's double-slit experiment in 1801 showed how light demonstrates the properties of both a wave form and particles.

This was considered impossible at the time, according to the classical view of physics. The experiment was showered with scorn and hostility because some thought it contradicted Newton's particle theory of light. Critics at the time argued that Young's publication 'contains nothing which deserves the name, either of experiment or discovery, and ... is destitute of merit ... We wish to raise our voice against innovations, that can have no other effect than to check the progress of science, and renew all those phantoms of imagination which ... Newton put to flight from her temple.'[33]

Thomas Young (1773–1829), British scientist.

Time has not been kind to these critics. In 2002[34] the experiment was voted as the most beautiful experiment in physics, and it has revolutionised humanity's understanding of reality. Again we see what happens when we hold on to first principles too tightly.

In the basic version of this experiment, a light source, such as a laser beam, illuminates a plate via two parallel slits.

The light passing through the slits is observed on a screen beyond the plate. The wave nature of light causes the light waves passing through the two slits to interfere, producing a pattern of bright and dark bands on the screen that would not be expected if light consisted solely of particles. This is because waves maintain a connection through the slits and bend as a result of the slits. They recombine on the other side but are distorted as a result of the slits. Peaks and troughs cancel each other out and combine to create larger waves. The interaction creates the 'interference' pattern that typifies wave form dynamics.

It is different from the pattern than would have been created if light was made up of discrete independent units (or particles), as these units would not need to have their trajectory altered because of the movement of another unit (as they are not connected).

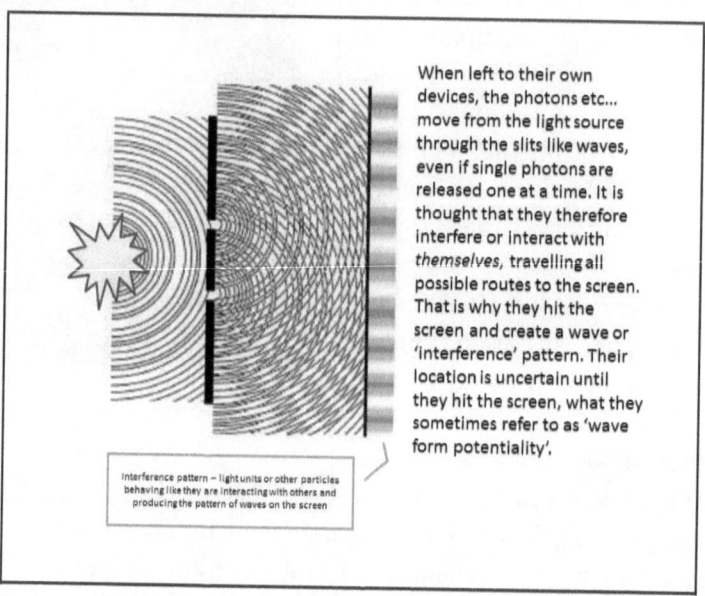

When left to their own devices, the photons etc... move from the light source through the slits like waves, even if single photons are released one at a time. It is thought that they therefore interfere or interact with *themselves*, travelling all possible routes to the screen. That is why they hit the screen and create a wave or 'interference' pattern. Their location is uncertain until they hit the screen, what they sometimes refer to as 'wave form potentiality'.

Interference pattern – light units or other particles behaving like they are interacting with others and producing the pattern of waves on the screen

The curious thing about light is that it has the ability to act like both a wave form *and* as if it was made up of particles. Like electrons, photons (small particles or 'packets' of light which are the components of light) can be sent out individually from a source. So the question is, if we reduce the number of electrons or photons in the beam to, say, one per second, does the interference pattern disappear?[35] After all, if photons were sent on their way one at a time, there would be no other photons for them to bump into.

Astonishingly, the answer is no. We see the individual photons and electrons strike the screen, and with time the interference pattern builds up, as if a wave form was sent through the slits rather than single units. Notice that with such a slow rate, each photon (or electron) is *not* interacting with other photons to produce the interference pattern. This is because each photon or electron is interfering with *itself*.

The formation of the interference pattern requires two slits, but how can a single photon passing through one slit *know* about the existence of the other slit? That is, how does it *know* that its trajectory must change to accommodate an interference pattern on the screen as a result of there being another slit for it to potentially go through?

The only solution is to give up the idea of a photon or an electron having location. The location of a subatomic particle is not defined. Scientists believe the particle *goes through both slits*.

Modern science has been able to show, through numerous extensions of Young's experiment, that these results can be replicated for larger objects, from electrons, to protons to *buckyballs*[36] (microscopic 'footballs' made of carbon atoms).

The only time that light behaves like a true particle instead of a wave form in this scenario is when something else is introduced to the environment, that is, *the observer,* aka you and me.

ROLE OF THE OBSERVER

So what is the role of the observer?

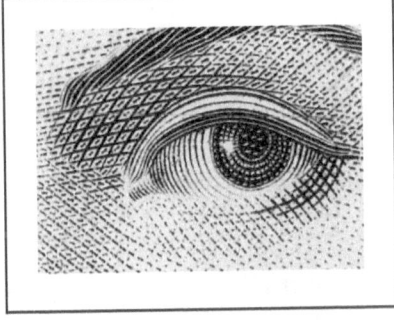

As the quantum world cannot be perceived directly, we need instruments to see what is happening. The problem is that the act of measuring appears to disturb the energy and position of subatomic particles. This is called the *measurement problem.*

One way to see the observer effect in the wave and particle nature of the quantum world is to return to the two-slit experiment and try to determine which slit the photon goes through. If the photon is a particle, it must go through one or the other slit; if a wave form, it goes through both.

Testing shows that by us *looking* at a point in one of the slits the interference pattern is wiped out and the wave nature of the light is eliminated. Only the particle nature remains, thus producing no interference patterns. The pattern which appears on the screen resembles what one would expect if one shot a lot of tennis balls through two slits in a wall. They do not produce the pattern that would be expected if the area was flooded with water, and waves rolled through the two slits in the wall (that is, the wave form interference pattern).

It appears that observing the behaviour of particles has a role in bringing them into material existence. Until we observe them, they are potentially anywhere in the universe in *superposition*, existing as a wave form of potentiality (what renowned quantum physicist Richard Feynman described as the *sum of histories*[37]). After we apply our observation (measure them), they become matter and *collapse* into our everyday, material world.[38]

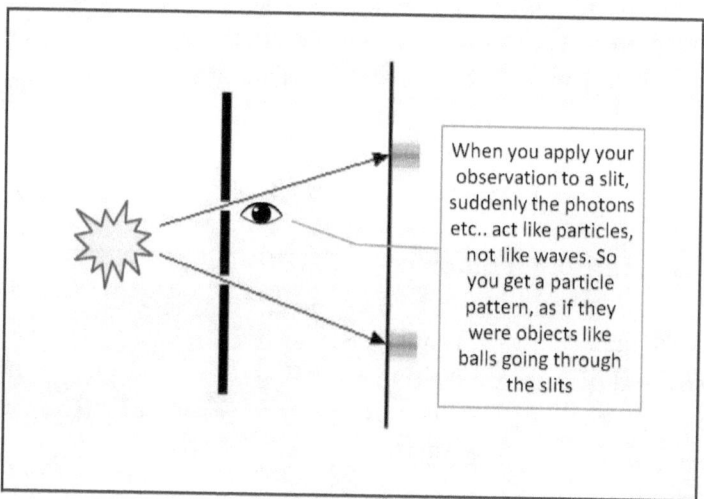

When you apply your observation to a slit, suddenly the photons etc.. act like particles, not like waves. So you get a particle pattern, as if they were objects like balls going through the slits

Variations of this experiment have shown that not only can objects be entangled in different locations, but they can be entangled in different times (the *Delayed-Choice Experiments*[39]). A famous extension of the delayed-choice experiments was first performed in 1999 by Kim et al.,[40] called the *Delayed-Choice Quantum Eraser*. These experiments have confirmed that the equipment used in the testing is not causing the effect and that information can actually be communicated *back in time*. It also confirms that knowing something subjectively supports the existence of our material reality.

Despite the fancy terminology, the basics of the experiments are not difficult to understand. What is hard to understand are the results. But take heart, no one can truly understand them, or is even claiming to. Following, I will describe the experiments so that you can expose your mind to one of our most profound discoveries about the universe we live in.

Delayed-choice experiments[41]

As discussed above, in Thomas Young's double-slit experiment, light is separated into two slits. Another way of demonstrating the phenomenon is to subject a photon to a beam splitter, which randomly directs photons one way or another, with each path being geographically different, preventing one photon from interfering with another.

In the diagrams, photons are emitted, one at a time, from the bottom left. They each pass through a 50 per cent beam splitter which reflects 50 per cent of the photons one way and 50 per cent the other way.

The top diagram shows the paths. One path goes north, then east, and passes the intersection on the top right-hand corner. It then continues east.

The other goes east, then north, and continues north past the intersection. In these cases, as the trajectory of the paths can be known (by an observer[xvii]), the result is that the photons collapse into material particles and register as such on the eastern and northern sides of the top-right intersection.

---

xvii Note that the act of observation separates. When you observe something, you are disassociating it from yourself.

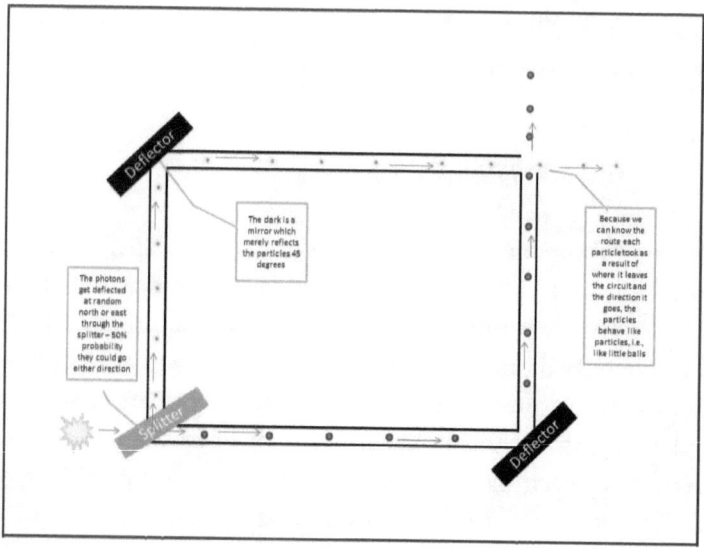

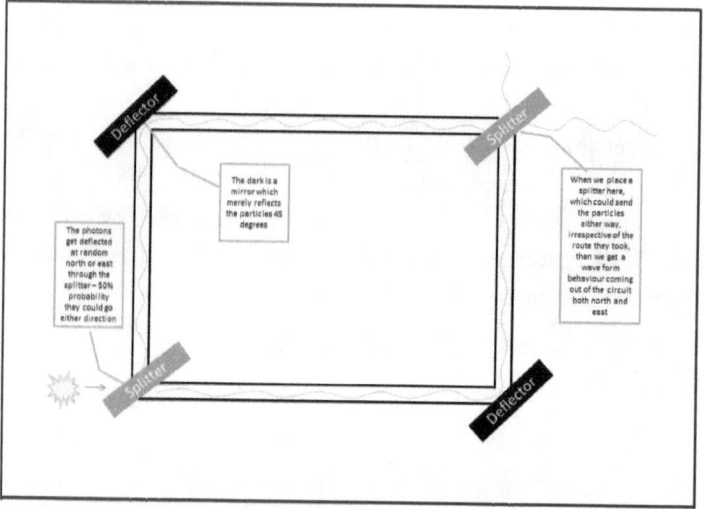

In the second diagram, a second beam splitter is introduced at the intersection at top-right. It can direct either beam towards either path. Whichever emerges from either direction could have come from either path. Its trajectory (via the top or bottom path) is unknown. In this way its trajectory information is said to have been *erased*.

Tests have shown this to produce, on the north and east of the intersection, an interference pattern, even if one photon at a time is released at the start. It means that instead of going down one path or the other, the photon actually goes via *both* pathways, interfering with itself, causing an interference pattern on the other side of the top-right intersection.

This test has shown experimentally, that the *decision* to travel the circuit as either a particle or a wave form (travel via both routes—wave form potentiality) is made at a time after it enters the circuit.

In the second diagram, light (at the speed of light) reaches the top-right intersection and finds another splitter which can randomly redirect the photon in either direction (east or north). It 'realises' that its trajectory becomes unknowable as a result of the extra splitter, and then signals back in time to the place it started from, to travel through the circuit as a wave form instead of a particle.

The only real difference between the first and second scenario, is our (yours and mine) subjective conscious ability to *know* the trajectory. It also shows that information can be relayed *back in time* to change the future; that is, when confronted with the second splitter, information is relayed back in time to the entry point in the circuit, to signal the need for the photon to move through the circuit not as a particle but as a wave form; that is, to take both routes and interfere with each other (or itself, for that matter).

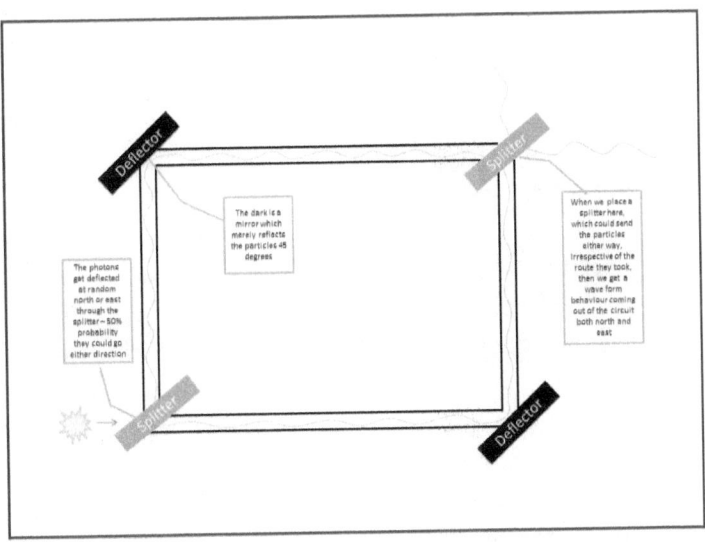

Many of you would quite rightly suggest that perhaps this is just the result of the equipment. Something about the equipment has changed the results, rather than our subjective knowing. In another famous experiment, called the delayed-choice quantum eraser, by Kim et al.,[42] this theory was tested. The results were clear. The equipment is not the cause of the change, and it is indeed our knowing, or our ability to *know*, that is the changing variable. The experiment is a little more complex, and for some a challenge to follow, but with perseverance most will understand it. It took me a while to understand it—until I came across a couple of explanations that worked for me. It helps to have it explained in more than one way, and I have put two explanations that I have found useful in Appendix 1 for those interested in the science.

If you are having difficulty understanding the implications of the experiment, consider it this way: imagine for a moment that you represented the sum total of consciousness in the universe. That you were it. OK, it might be hard to imagine, but work with me for a bit. Now imagine that you walked into a room and, after you closed the door, the room blocked off your senses to the outside world—you could not hear, see or smell anything beyond the room. The above experiment suggests that the world that you know, which you saw and experienced before you closed the door, would no longer exist in the state you left it. Your ability to perceive the world around you is integral to it being able to exist in the form it does. Without your perceiving it, it would be just wave form potentiality (whatever that would look like). The world as you see it does not exist independently of you, without consciousness being able to observe it.

The important thing at this point is to understand that the discovered role of the observer has monumental significance in modern science, for it introduces the agency of something which traditional science rejects as irrelevant to reality and the mechanics of the natural world, that is, the *subjective*. In this context the subjective is considered to be the nature of the mind—our consciousness, our ability to know—compared with the *objective* material thing we observe. The two-slit experiment, for the first time in physics, indicates that there is a much deeper relationship between the observer and the phenomenon, at least at the subatomic level. This is an extreme break from the idea of an objective reality that exists without requiring us to perceive it, which would exist even if we did not.

In fact, many classical theorists, and their modern counterparts (as we shall discuss in more detail later), argue that the subjective is even less than irrelevant; that is, that it doesn't actually exist. As you might imagine, the effect of these experiments being done again and again, by different respected scientists around the world and producing the same results, significantly undermines the classical position.

The only way to understand these results is to acknowledge that firstly, there is a realm of reality, a 'subspace' that is not bound by space-time as we know it; and secondly, that consciousness, the phenomenon of knowing something about the world beyond the action–reaction mechanics of space-time, is intimately connected to collapsing energy from potentiality (wave form) into matter (particles).

In fact, quantum physics shows us that our conscious observation does more than deliver material particles; it also delivers the history of the universe. As Stephen Hawking et al. suggest, 'quantum physics tells us that no matter how thorough our observation of the present, the (observable) past, like the future, is indefinite and exists only as a spectrum of possibilities. The universe, according to quantum physics, has no single past, or history… we create history by our observation, rather than history creating us.'[43]

The founders of quantum theory—including Neils Bohr, Max Planck, Werner Heisenberg, Erwin Schrödinger and Albert Einstein—wrote extensively about the implications of quantum physics, aware that the role of the observer was a radical change in how physics had been practised.

'I regard consciousness as fundamental. I regard matter as derivative from consciousness.' Max Planck is considered the father of quantum theory. German physicists Max Planck (c. 1930).

Some physicists – including Wolfgang Pauli, Pascual Jordan and Eugene Wigner – believed that consciousness was not merely important but was fundamentally responsible for the formation of reality. Jordan wrote, 'Observations not only disturb what has been measured, they produce it ... We compel [the electron] to assume a definite position ... We ourselves produce the results of measurement.' This is known as the *strong view* of the role of consciousness and is also supported by renowned physicists John von Neumann, Bernard d'Espagnat, Euan Squires and Henry Stapp.[44]

Some think it was the eighteenth century philosopher George Berkeley who inspired the thought experiment of whether a tree falling in the forest made a sound if there wasn't anyone around to hear it. It is similar to the question put to Niels Bohr, one of the founders of quantum mechanics, by Albert Einstein; that is, whether the moon exists if nobody is looking at it. Perhaps we are getting closer to the answer, but are we ready for it?

ANTHROPIC PRINCIPLE

Shortly we will be moving from the physics part of *the elephant* to the biological part. Moving from nuts and bolts to living beings like us is not as easy to explain as some might think.

This jump from mechanics to life—and, indeed, then to beings like us—has baffled thinkers for thousands of years, and even today the more sober scientists are still uncertain as to how much closer we are to real answers.

As we will see in the next section, life itself provides us with a unique and strong perspective on a possible subspace. But to firstly understand the perplexing question of life, it is helpful to understand the dialogue in physics around the *Anthropic Principle*. This refers to the remarkable coincidences that have come together to create a condition that supports our existence as living beings.

These coincidences include the 'luck' of having a solar system able to have a *Goldilocks Zone* (a radius of rotation around a sun which is not too hot or cold), a uniquely suited planet like Earth orbiting in the zone, just the right kind of sun with just the right kind of radiation, with a Jupiter-like planet around to draw menacing meteors away from Earth, and a position in the galaxy out of the action enough to keep us safe but also in a unique location to be able to observe the universe. It also includes, among other things, the fine-tuning of the universal constants, such as the forces (weak and strong nuclear, electromagnetic and gravity) the mass of the proton, the number of dimensions, and the setting of the cosmological constant (that force which prevents the universe from imploding).[45]

The probability of a universe like ours existing, coupled with the probability that life would develop, and then develop into creatures like us, is so minuscule that it is probably as close as we can come to impossible, without actually reaching it. As the evolutionary biologist Richard Dawkins says,[46]

'... the origin of the eukaryotic cell (our kind of cell, with mitochondria, which are not present in bacteria) was an even more momentous, difficult and statistically improbable step than the origin of life. The origin of consciousness might be another major gap whose bridge was of the same order of improbability.'

Which is good for us, because if we reached impossible—we would not exist. Our existence probably has a place in the cosmos hall of fame as proof positive that anything is possible. As Hawking put it, 'our universe and its laws appear to have a design that both is tailor-made to support us and, if we are to exist, leaves little room for alteration. That is not easily explained, and raises the natural question of why it is that way' (Hawking et al.[47]).

Some scholars have attempted to explain this fine-tuning with a theory proposing an endless number of universes, each with its own natural laws and constants; which makes our chances of existing not so special. For example, say that the chance of our Earth and universe existing is one in a trillion. This theory suggests that as the number of universes apart from our own is infinite, a trillion would fit into infinity so many times that the chance that our Earth and universe could exist goes from improbable to probable, to incredibly likely right through to its existing an infinite number of times.[48]

Max Tegmark, a cosmologist and physicist at MIT, says this can all happen in *this* universe, because there is no evidence that that space is finite. He proposes that if one travels far enough (he suggests 10 to the power of 10^{29} metres), one will meet an identical copy of themselves. He also suggests that our nearest universe twin is 10 to the power of 10^{118} metres away.[49]

As you may gather, there are problems with this theory. First, the birth of many universes also creates a greater universe (or *multiverse* within which the universes exist). The chances therefore that not only our universe has the fine-tuning it has, but also that the greater universe has its own (in order to create the many universes required to produce our own) reduces the probability even more dramatically. So this solution isn't really a solution at all, as it brings us back to the same question, how do we explain the fine-tuning?

Perhaps a bigger problem with the theory relates to the belief that, scientifically, nothing is absolutely impossible, a fact supported somewhat by quantum physics. A live tyrannosaurus wearing a digital watch might seem impossible but, in absolute terms, it has a probability, as minute as it may be. Say it has a probability of one in a million billion trillion. Now, since a million billion trillion fits into infinity an infinite number of times (according to the above rationale) a digital-watch-wearing T-Rex would not only exist, but is likely to exist ... ultimately ... an infinite number of times.

Moreover, as the theory was partly developed to show how the universe could have come about without a creator to help it along,[xviii] the logical extension of the above is the proposition that there is a probability associated with the existence of a creator. Let's suggest that the probability of a creator existing is one chance in a trillion tredecillion (which has 54 zeros).

As a trillion tredecillion fits into infinity an infinite number of times, a creator therefore is likely to exist an infinite number of times. However, as there can be only one ultimate creator (by definition), the probability may well collapse into the existence of a creator. So this theory really leads us nowhere. Or does it?

Probability would have a place for pretty much anything, including a live T-Rex wearing a digital-watch. Source: myfavoritedinosaur.com but altered. Digital watch added.

Tegmark and others also suggest that there is a level (level 4^{50}) platform of universes which include advanced civilisations that create advanced simulations in which *we* are players. It is suggested that because these civilisations are so advanced, we would be unable to perceive that we were, in fact, in a simulation.

Some cosmologists (string theorists, in particular) suggest that you don't need an infinite number of multiverses to produce one like ours, perhaps only 10^{500} (that is, 1 followed by 500 zeros). However, in the

---

xviii '…in the vastness of astronomical space, or geological time, events that seem impossible in Middle World [our everyday world] turn out to be inevitable. Science flings open the narrow window through which we are accustomed to viewing the spectrum of possibilities. We are liberated by calculation and reason to visit regions of possibility that had only once seemed out of bounds or inhabited by dragons.' *The God Delusion*, Dawkins, 2006, p. 419.

words of the renowned Cambridge University professor of mathematical physics, and theologian, Sir John Polkinghorne, 'if you are allowed to posit 10^{500} other universes to explain away otherwise inconvenient observations, you can "explain away" anything, and science becomes impossible'.[51]

Is there a way to explain this phenomenon of life? Some scientists suggest evolution explains it all. Others suggest that looking at the question only from a space-time point of view will never provide us with the complete answer. Let's now look at the evidence for subspace from our study of living things, but before we do, consider one interesting idea, which may shed some light on the above enigma.

THE UNIVERSE—AS A QUANTUM STATE

Princeton physicist John Wheeler proposed an interesting possibility. We know from quantum physics that it is possible for a particle of light to travel to us from a galaxy millions of light years away in a quantum potential state (see above discussion[52]) until the moment it reaches us (the conscious observer).

At this point it becomes an event, and enters our reality as a material object (particle).[53] Its history as a particle is created at that point, going back millions of years. Wheeler proposed that this could mean that the universe is brought into reality retrospectively in a similar way, through our observation.[54]

Using the example of the delayed-choice experiments, imagine a scenario in which the particle's route is 'erased'. Imagine that this is the scenario representing the absence of an observer. As the course is not known, it travels to Earth on a journey that lasts 100 million years. As the course is not known until a conscious observer (such as us) perceives it, it arrives as a wave form to our planet.

PHYSICS—THE LEGS

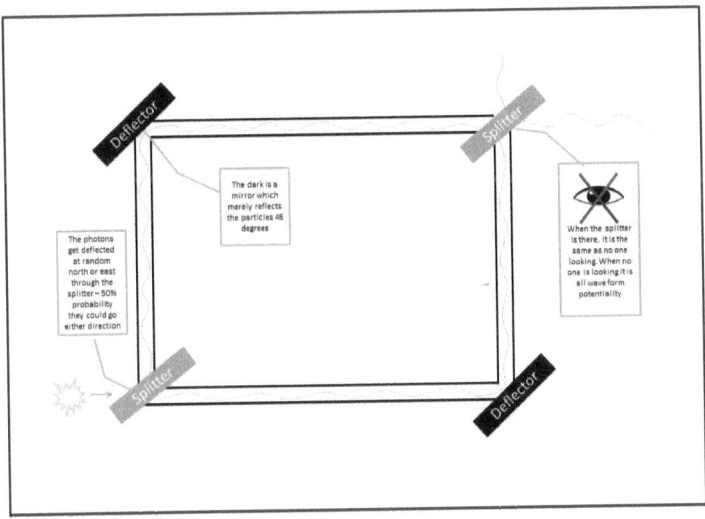

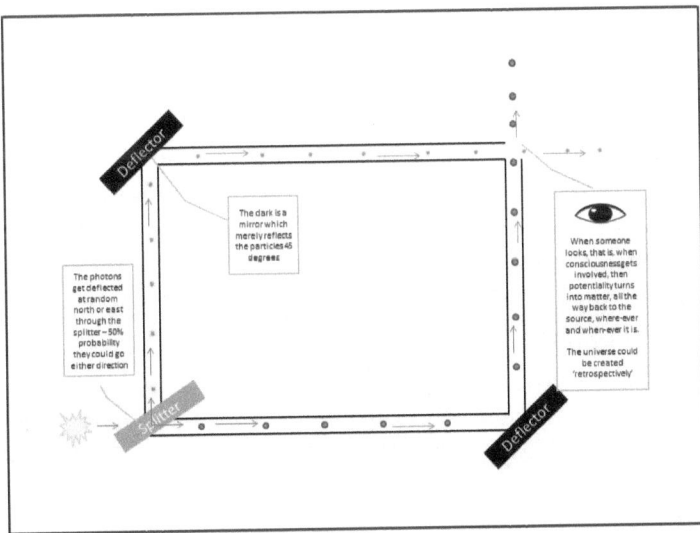

The second diagram represents what happens when there is an observer. The conscious observer is introduced by taking away the second beam splitter. At this point, information travels back 100 million light years so that the photon can affect us as a particle, with defined properties as matter, rather than a wave form of potentiality.

The information goes back in time to say, 'Hey, we are being watched. Quick, change into a particle and let's do this again, this time by actually travelling the 100 million light years as a bit of matter.'

Would this mean halting everything for 100 million years before giving us the particle? This is doubtful. It is more likely that the information is communicated back in time, so that *history is changed*. Before we observed it, the matter didn't exist in material form, only as potentiality. After we observe it, it becomes matter, even from its point of origin.

This theory suggests that the universe was able to come into being with the first observer, collapsing the potentiality of all possible universes (for example, all possible universes in superposition) into our actual universe—the only universe able to sustain our material existence.

In this scenario the *first cause* is immaterial consciousness, with information.

These ideas are being explored today by scientists who are looking at practical approaches to quantum physics. The renowned physicist Anton Zeilinger provides the example of Jorge Luis Borges's story about a library which has all books in it, including those which have been written and those which are yet to be written.[55] It is theoretically possible to have all books which will be written, because it is merely an exercise in creating all possible combinations of letters and words (similar to the *infinite monkey theorem*[xix]). But such a library would, in practice, be impossible to realise, let alone navigate.

Zeilinger suggests that quantum computers are being developed in classified laboratories around the world.[56] It is possible that quantum computers can work off this principle and virtually collapse all possible potential books which are not the books about the secrets of the universe. What is left, suggests Zeilinger's approach, might be the book on the secrets of the universe.

---

xix *The infinite monkey theorem* states that a monkey hitting keys at random on a typewriter keyboard for an infinite amount of time will almost surely type a given text, such as the complete works of William Shakespeare.

SUMMARY

We have seen that the material world so familiar to us is all but an illusion; that is, the world of solid objects, of matter, is really a representation of an existence which is not solid at all. Matter is merely a manifestation of it.

We also see that space (distance) and time are also phenomena manufactured out of a more primal subspace-like dimension. This subspace-like reality, in which time and space do not present the same barriers, reveals itself in signs of entanglement, objects being connected even though they are separated by space and time. It also reveals itself in quantum experiments showing information travelling back in time, demonstrating how the future has the ability to constrain the past.

A far greater mystery is our ability to share an experience of now, to perceive a clump of time at once, which travels in one direction of causality. There appears to be no accounting for this phenomenon in the physics of space-time reality.

Physics has also become aware of some tell-tale seams in creation, revealing the role the mind has in bringing matter (and perhaps our reality) into being out of a sea of possibilities. It points to a possible involvement of consciousness in the physics of reality. It is hard to overplay this development.

We also have the mystery of the fine-tuning of our universe and the synchronicity of events which have come together to create life. In order to make sense of this and other phenomena without relying on a divinity as a possible agent in the universe, physicists have modelled a universe with many other dimensions than those we are familiar with, and indeed many—if not an infinite number of—other universes. Yet when we test the models, they still fall short as explanations, as they fall victim to the very problems they were designed to solve. Could Wheeler's suggestion, that consciousness may have created the universe retrospectively, provide a solution?

CHAPTER 3
Biology, the body of the elephant

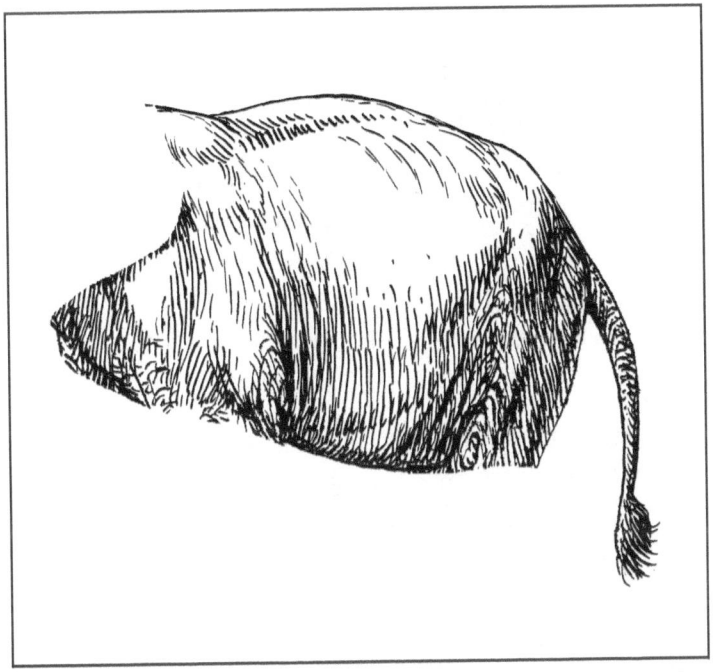

An online dictionary defines *biology* as the science of life or living matter in all its form and phenomena, especially with reference to origin, growth, reproduction, structure, and behaviour.[1] What does a review of scientific thinking and research in biology show us about subspace? First we have to appreciate that the technology of the living world is far, far more advanced than anything humans have created or will create for some time to come.

THE GREATEST COMPUTER IN THE UNIVERSE

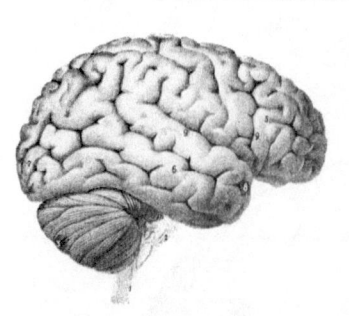

Is our brain really more like a quantum computer than a conventional one? Outer surface of the human brain 1894–1895. Popular Science Monthly Volume 46.

About half a million years ago our neocortex reached the size it is today, providing humans with special abilities and a unique perspective on reality. This included a self-conscious mind 'awake' to its environment, arguably to a greater extent than that of other creatures.

Our unique brain has given us an advantage over the environment and other animals. While modern research tells us that a lot of our behaviour originates subconsciously,[2] there is controversy around the role of our conscious (or awake) mind and how it mediates and manages the messages it receives from its subconscious counterpart.

Many consider consciousness to be the ultimate example of something that is self-evident, like our ability to see, hear or even just know we exist. It might be difficult to explain in objective, quantifiable terms, but if you have it, you can't miss it. If you don't have it, you have absolutely no idea what everyone else is talking about, no doubt thinking that everyone else is experiencing some kind of mass delusion.

It is like trying to explain what hearing is like to someone who has never heard a sound; or to a person who has never had the sense of sight, the difference between red and blue. There is no clear point of reference. The only way to show that some of us can in fact see and hear is to show its effects, the abilities or phenomena that can't be explained without these capabilities or capacities.

In the pages ahead, evidence will be presented for the existence of consciousness; that is, abilities and phenomena that together can't be plausibly explained without the existence of consciousness.

As we will discover, our brain has been able to sustain a functioning consciousness, which together with our subconscious may be able to provide important information about the subspace realm, a realm that traverses space and time. Indeed, as we have seen above, contemporary physics tells us that our ability to deliver conscious observation to our environment has an important role in bringing matter into reality out of a sea of possibilities.

For those of us who possess consciousness, we experience a point in time. Our daily lives are made up of mediating the messages we get from our subconscious. These can be instinctual drives, like hunger and thirst, but also intuitive knowledge arising out of the brain's

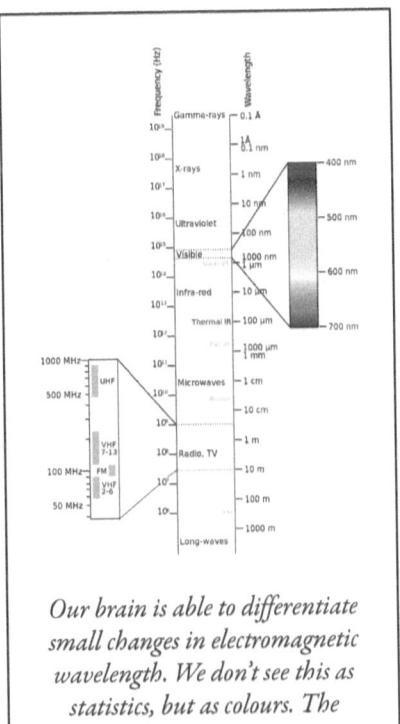

Our brain is able to differentiate small changes in electromagnetic wavelength. We don't see this as statistics, but as colours. The electromagnetic spectrum. Permission granted by Victor Blacus.

ability to detect highly complex patterns in wavelengths and frequencies received through the senses, from colours (electromagnetic-wave spectrum) to different musical notes (mechanical wave forms).

Our capacity to identify hyper-complex patterns doesn't stop at colours and musical notes. It goes much further with our capacity to appreciate 'higher order' phenomena, such as stories, music, comedy, meaning and spirituality. These are examples of phenomena too complex for our conscious minds to fully grasp.

Intuition appears to act as a pathway between the conscious and subconscious, providing our consciousness with information in a form it can use. Without this interface the hyper-complex patterns would overwhelm our minds with a mass of data and computations, making it difficult for us to manage. In this way, our self-aware consciousness is probably a little like the random access memory (RAM) in a computer.

The partnership between our conscious and subconscious is plainly a highly successful one allowing us to achieve feats far beyond the capacity of the most advanced computers, like driving a car while planning dinner.

Neuroscience has acknowledged the important role of feelings, how they are intimately related to reason and how feelings and reason are not opposites.[3] An experienced mechanic is able to *feel* something not quite right in the sound of an engine before becoming aware of exactly what the problem is. A mother can often sense that there is something wrong with her child before identifying the problem. A talent scout will intuitively spot a young entertainer who's *got it* long before perceiving exactly what *it* is. A skilled musician will be able to identify a wrong chord in an orchestra before knowing which instrument is to blame. How often have you seen a movie and thought 'wow, that was great' while not necessarily being able to explain, in any objective sense, why it was so special?

Without such an interface, and an ability of our consciousness to orchestrate our subconscious powerhouse, humans would not be able to negotiate the complexities of their existence and would likely still be living in trees.

With another of the neocortex's gifts, complex language and writing, we have used the scientific method to unravel intuition and create a common, manifest knowledge. This takes our individual efforts and creates libraries of accumulated knowledge for us to build on. With each generation, more manifest knowledge is collected and transferred through our educational institutions into the minds of the next generation. Doing this has enabled us to collectively accomplish far more than we could as individuals, from stone spearheads a half a million years ago to a self-guided artificially intelligent probe roaming the surface of Mars today.

While we have farmed our knowledge successfully in some areas, such as engineering, we have been less successful unravelling our intuitive knowledge in other areas.

For a lot of subjects this does not matter so much, thanks to our resident geniuses. Take for instance our ability to appreciate music, storytelling and comedy. Ludwig van Beethoven's *Ode to Joy* is a piece of music appreciated the world over which, sadly, Beethoven himself never heard, due to his deafness at the time of writing it. We can all appreciate its merits even though most of us would not be able to explain, in any 'scientific' or mechanical sense, why the piece is so special.

The same can be said for storytelling, movie-making and comedy. Indeed, if there was any certainty around the science of these, the critics would be making masterpieces themselves. Or perhaps computers would be writing all the movie scripts and replacing stand-up comics at comedy clubs.

Thanks to our own intuitive abilities and the talents of people like Mozart, McCartney, Williams, Spielberg, Shakespeare and Seinfeld, we don't have to consciously understand the science behind it. The intuitive knowledge we gain from our super subconscious allows us to detect and appreciate quality with confidence, and we don't need (or even want) any 'expert' to tell us what we should or shouldn't appreciate.

A scientist would have a hard time trying to sell a mathematical proof that Bach's *Air on the G String* is rubbish, or that *Seinfeld*

We don't have to look back in time for geniuses. Like Mozart, John Williams writes for the common ear. He wrote some of the most iconic music of our time, such as the music for the movies STAR WARS, ET, JAWS, JURASSIC PARK, HARRY POTTER, INDIANA JONES, SAVING PRIVATE RYAN, SCHINDLER'S LIST, SUPERMAN, CLOSE ENCOUNTERS OF THE THIRD KIND, *and series such as* LOST IN SPACE *and* LAND OF THE GIANTS. *He has won five* ACADEMY AWARDS. *Williams at the Boston Symphony Hall after conducting the Boston Pops, May 2006. Source Author Nationalparks.*

has no humour in it, and that only deluded fools would think otherwise.

However, our intuition tells us other things arguably more important than whether something is funny or acoustically synthesised. Throughout our history, we have demonstrated another of the neocortex's endowments: a strong, continuous, consistent and persistent intuitive knowledge that there is more to existence than bricks and mortar, long before quantum physics claimed the discovery.

From the Pyramids to the White House, our intuitive perception of what quantum physics is now telling us—that is, of a wider, metaphysical realm—has played a major role in defining us as a species and the building of civilisation as we know it today.

We are going to examine within biology how the most powerful computer in the world can contribute to our knowledge of a subspace realm.

If we accept the broad definition of biology, such as the one above, the study of our mind and brain needs to be included. As we shall see, *consciousness* does provide some remarkable insights, and modern science is becoming better at grappling with it, measuring it and understanding this unique phenomenon.

PARADIGM WARS

Getting behind the phenomenon of consciousness has not been easy. Historically, consciousness represented a phenomenon that contradicted the way the world ought to be seen. It fanned political feuding between scientific paradigms that continues to this day.

The nineteenth century, *naturalist,*[xx] view of the universe found consciousness difficult to deal with. This was largely because it violated the bottom-up view of the mechanics of the universe; that is, everything started at the Big Bang and everything since has been strictly the fall of dominoes, in accordance with the so-called 'natural laws', such as

---

xx Sometimes the naturalist paradigm is also referred to as *materialism, mechanistic science*, or the *classical* or *Newtonian* view of science/physics.

gravity. According to this view there is no rhyme or reason about existence, or freedom to go beyond the natural laws or domino effect.

This position is captured by the mathematician and philosopher Bertrand Russell.[4]

> *'That man is the product of causes which had no pre-vision of the end they were achieving; that his origin, his growth, his hopes and fears, his loves and his beliefs, are but the outcome of accidental collocations of atoms; that no fire, no heroism, no intensity of thought and feeling, can preserve an individual's life beyond the grave; that all the labours of the ages, all the devotion, all the inspiration, all the noonday brightness of human genius, are destined to extinction in the vast death of the solar system, and that the whole temple of Man's achievement must inevitably be buried beneath the debris of a universe in ruins—all these things, if not quite beyond dispute, are yet so nearly certain, that no philosophy which rejects them can hope to stand. Only within the scaffolding of these truths, only on the firm foundation of unyielding despair, can the soul's habitation henceforth be safely built.'* In *A Free Man's Worship*, by Bertrand Russell, 1923, 6–7.

Not all scholars of the naturalist age agreed though. Charles Darwin, for example, did not give up on meaning or a higher human purpose. As we shall see later, people like Darwin believed that a God may very well have created the universe and all life,[xxi] including each of us, at the Big Bang. I suppose for an all-powerful omnipresent and timeless entity perched in subspace, watching 13 billion years go by since the Big Bang would be a little like us watching pasta sauce simmer for 20 minutes.

Although naturalism is curiously silent about who or what enforces the laws (which for many is a key question), it maintains that the natural laws facilitate an action-reaction process, which is all that exists. It is proposed that this mechanical domino effect cannot be interfered with in any way, not even a very little tiny bit. If anything does, even

---

xxi People who have these beliefs are referred to as *deists*. More later.

once, the paradigm must fall. It is therefore puzzling that Russell titled his book *A Free Man's Worship*, when his position suggests humanity has no undetermined agency of its own, that is, no free will.

This view has also been referred to as a *reductionist* approach, as it proposes the reason or cause for everything as being the result of mechanical interactions from the micro to the macro, from the small to the big. Understand the micro and you understand the macro.

A case study: Phrenology was developed by a German physician, Franz Joseph Gall, in the late 1700s. He assigned bumps on the skull to certain personality types including criminality, by examining the heads of pickpockets and prisoners etc. With his young pickpockets, for example, he suggested that a bump behind the ear was associated with a tendency to steal, lie or deceive.

In his book on the subject of phrenology, Gall suggested that:[5]

1. Moral and intellectual faculties were innate.

2. The exercise or manifestation of these faculties depended upon their organisation.

3. The brain controlled all of the propensities, sentiments and faculties.

4. The brain was composed of as many organs as there were different faculties, propensities and sentiments.

5. The form of the skull represented and reflected the form and development of the brain organs.

Gall developed a system of 27 *faculties* which he believed could be directly diagnosed by assessing specific parts of the head. He created a chart showing which areas of the skull were associated with specific traits or characteristics, from loving one's offspring to vanity to religion.

Despite opposition from the church (mostly because he nominated a part of the brain responsible for religious thought), Gall's ideas struck a chord with naturalistic science. The field became increasingly popular through the nineteenth century and even to the early twentieth. The suggestion that biology determined human characteristics was consistent with the determined world which naturalism advocated.

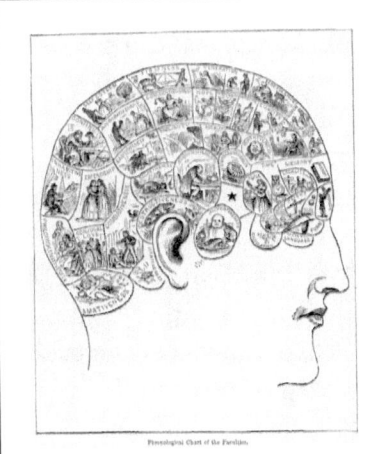

In the tradition of space-time explanations, the idea that behaviour could be explained principally by someone's inherited biology was a popular theory among nineteenth century naturalists. A phrenology chart from PEOPLE'S CYCLOPEDIA OF UNIVERSAL KNOWLEDGE *(1883).*

Though neuroscience today acknowledges that certain parts of the brain are associated with specific functions, phrenology, particularly the type linking skull shapes with criminal conduct, is no longer taken seriously.

During phrenology's popularity, however, someone who proposed that criminality might also be linked to social and environmental factors (noting the other commonalities widely acknowledged today) would likely have been ridiculed, as the concept did not fit within the phrenologists' scientific paradigm that the brain was the source of behaviour and all mental processes, that is, the above five premises. It just would not fit the prevailing philosophy.

This kind of logic may sound absurd, but the analogy can explain many of the conflicts that have occurred between a naturalistic, determined view of the universe and one which attempts to include consciousness and other observed phenomena. What is found in the real world of experiments is often judged not for its own merit, but for its value in contributing to the philosophical positions of established science at the time.

Over the last hundred or so years, scientific experiment and discovery (reviewed in the above sections) has led scientists and researchers to look for a paradigm that reflects new findings, accepting that the naturalist view of the world is incomplete.

While evolving our understanding can challenge some of our cultural views—in particular, those of some parts of the Western scientific establishment[6]—these new views are fostering an acceptance that there is more to immaterial reality than the so-called natural forces, imperceptible to the normal senses, not totally constrained by the space-time 'laws' of the 'natural' world.

As one might imagine, the absolute naturalist view has great difficulty surviving quantum physics. Quantum physics cannot support a material reality, let alone the so-called 'laws' of nature, because they lose their authority in quantum space. As we have seen in quantum physics, the laws of nature that we experience in our everyday world do not apply in the same way to the tiny pieces that make up that world. As Hawking suggests, '... components of all objects obey the laws of quantum physics ...' even though the natural laws are a guide to how microscopic matter behaves *en masse*.[7]

Contemporary scientific investigation is even questioning the assumption that reality has unbreakable laws. Laws are fundamentally human inventions. There are no rule books in the universe. As Sheldrake observed, the laws of natural science originated from a religious concept that God had laws to be followed, including laws over nature.[8]

Once more physicists tell us that, within the singularity that caused the Big Bang, energy was so ordered that our natural laws did not exist.[9]

Some scientists explain the fine-tuning of our universe with the suggestion of an infinite number of alternate universes, each with its own natural laws. In fact, researchers here on Earth are also aware of how our so-called unchanging constants (such as gravity and the speed of light) are changing in value over time.[10]

So the proposition that the cosmos is built on unchanging laws, or at least what we consider to be the natural laws, is as outdated as the horse and cart is to modern travel. This is not to say these 'laws' aren't useful. Nothing much would work if we didn't respect them. But they are no longer the universal absolutes they were once claimed to be, and we can't use them (at least on their own) to construct a view of absolute reality or existence, as some in the past have tried to do.

Discoveries in physics are suggesting that immaterial forces outside of the natural forces can interrupt the deterministic flow of the universe, just like gravity and radio waves work within the natural laws. As the delayed-choice quantum eraser experiments have shown, these forces include the role of our conscious observation in collapsing possibility waves into matter. It is therefore appropriate to our quest for subspace that we examine the effects and agency of consciousness, not only at a quantum level, but on our everyday level.

CONSCIOUSNESS

As discussed above, consciousness is probably the best example of something self-evident but difficult to explain.[11]

While some members of the scientific community consider consciousness an emerging property, coming out of the complexity of our brains, others consider it the primary element of existence, shining through our bodies like light shining through a film projector[12] (the brain and body being the lens and film in the projector).

What experiences the moment?

As discussed above, the laws of physics are fully reversible. There is no reason why the arrow of causation must go in the direction we experience it, and there is no need for us to experience only the here and now. Perhaps the explanation can be found in the existence of consciousness, a phenomenon that breaks, from its position outside space-time, the domino sequence of causation (that which would occur if time-space was left alone).

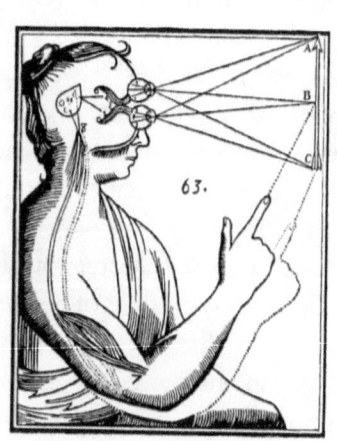

Illustration of mind and body dualism by René Descartes (1596–1650) in TREATISE ON MAN. *Inputs are passed by the sensory organs to the pineal gland and from there to the immaterial spirit.*

Consciousness may be a window into space-time from subspace.

Through the agency of consciousness there is a suggestion that we have some degree of choice, or discretion, about what happens in the material space-time world, creating more than one possible future. The idea is that consciousness is able to have agency in space-time, as it is 'powered' outside space-time dimensions.

It thus becomes necessary to clarify the role of our subjective existence, which numerous scientists have speculated about.

In line with some of the above findings, the quantum physicist Werner Heisenberg suggested that science 'no longer is in the position of observer of nature, but rather recognises itself as part of the interplay between man and nature. The scientific method ... changes and transforms its object: the procedure can no longer keep its distance from the object.'[13]

As Stephen Hawking commented above, our past, like the future, is indefinite and exists only as a spectrum of possibilities: 'The universe, according to quantum physics, has no single past, or history ... we create history by our observation, rather than history creating us.'[14]

It has been suggested that the agency of consciousness facilitates a different type of effect on our world, that is, a 'top-down' effect, driven by more complex contextual information. What is the evidence for this?

Is a top-down force theoretically possible?

We have seen above some support among physicists for the top-down agency. The role of the observer in the double-slit experiments and Stephen Hawking's musing on the creation of the past and the future through our observation, for example.

It also appears that mathematics may provide room for the phenomenon as well. Sir John Polkinghorne suggests that while reductionism can explain lower-level dynamics in systems, '... the behaviour of the system is conditioned by the configuration of the components ... there is a downward causation ... the system can only be understood in terms of higher-level properties'.[15]

Nobel Laureates John Eccles and Eugene Wigner support this view, with Wigner suggesting in his arguments about the symmetry of physics that the action of matter upon the mind must give rise to 'direct action of mind upon matter'.[16]

Interestingly, two of the most prominent and accomplished astrophysicists of the last century, Sir James Jeans and Sir Arthur Eddington, both espoused the view that consciousness was a likely foundation of the universe itself. Jeans wrote in *The Mysterious Universe*: 'The universe begins to look more like a great thought than like a great machine.'[17] As the physicist Henry Stapp put it, fundamentally quantum physics is 'a pragmatic scientific solution to the problem of the connection between mind and matter'.

The GESTALT DOG, *some people call it. The picture is an example of how a global perspective is essential in order to understand the significance of the black and white blotches. This famous photograph is of a Dalmatian dog on a beach.*

Is there evidence for the top-down effect?

Exhibit #1—human knowledge.

The classical view of reality considered free will an illusion,[18] but this position encounters difficulties when we consider our ability to generate *knowledge*.

In our daily life we use knowledge constantly, so much that we have become dull to its extraordinary nature. In order to gain objective knowledge—that is, knowledge that applies to a wide variety of situations—scientists need freedom to 'rise above' the natural, four-dimensional world of space-time, so that they can make observations of it and test these observations through predictions.

Picture a pool table covered with balls. The balls hitting each other represent the mechanics of the universe, 'governed' by action and reaction. One ball hits another. The resulting behaviour of the second ball is determined by the speed and angle of impact of the first. As one ball collides with a group of balls, a complex web of action and reaction is created.

Now imagine a pool table as large as the galaxy, with billions and billions of balls on it. Imagine the table is in a vacuum and constructed so that the momentum of one ball is conserved through trillions of interactions on the table. A person propelling a ball into such a table would create a monumentally complex web of actions and reactions, ultimately stretching across the galaxy.

There is a school of thought in science which suggests this complexity can reach a level that produces an illusion of self-awareness within the table, or at least within the collection of balls. It suggests that while the phenomenon of self-awareness exists, it is an illusion and has no agency in the world.[19] It would lack the capacity to change the action-reaction laws of physics which govern the connections each ball makes at the micro, local level.

For some, this view explains our own self-awareness or consciousness in a similar way. Its advocates, including philosopher Daniel Dennett and psychologist Susan Blackmore,[20] claim they have no consciousness at all, at least not one capable of affecting their behaviour. They also think that no one else has consciousness either. For them, the vast majority of people in the world who think they have consciousness are suffering from an illusion which, luckily, these two individuals are immune to. Other views are less extreme, suggesting that there may be a point at which the complexity creates a self-regulating entity that can produce what people experience as consciousness.

Sir Roger Penrose, an Oxford University mathematician and physicist, does not accept the complexity explanation for consciousness. He claims that computer algorithms, no matter how complex, are incapable of simulating mathematical reasoning and can therefore never produce consciousness. Penrose et al. have proposed quantum mechanical theories on the mind-brain relationship which suggest that consciousness cannot be localised to the brain. For him, the brain can facilitate, but not cause, the subjective experience of reality.[21]

Charles Scott Sherrington and John Eccles, both neuroscientists and Nobel Prize laureates, and the neurosurgeon Wilder Penfield were also of the opinion that the brain appears to behave more like a complicated organism that registers and transmits consciousness, rather than one that produces it. Indeed, in a recent book the neuroscientist Mario Beauregard demonstrates that a material approach to the mind-brain relationship is no longer tenable in neuroscience.[22]

Of course, the view of consciousness as just an illusion can appear to have the fatal flaw of being self-refuting; suggesting the existence of an awareness capable of experiencing an illusion. Also, how do we know that the illusion isn't itself an illusion?

It may also be possible, as we have suggested above, that there are some, like Dennett, who may not actually possess consciousness. Perhaps it is an error to consider that everyone has it just because some do. Trying to convince them of their consciousness would be like trying to convince someone who is colour-blind that they can see colours.

To convince them, we would have to show how the ability is of practical use, or provide examples of phenomena that can't be plausibly explained in any other way. Yet, as we discussed in the first section using the Ames room example, even when we point out the truth, people's brains still may not allow them to see it for what it really is.

Using the analogy of a galactic pool table, imagine that you are one of the balls. Picture yourself bouncing off one ball and then another.

You would be so affected by the immediate dynamics around you that predicting what would happen to you next would be virtually impossible. You would not be able to see how a ball got to you the way it did, or the sequence of events that led it to bump into you. And even if you did, you could not change anything, because you could react only to those balls hitting you. You would be just another bouncing ball, an impotent slave of your environment, as suggested by Dennett and Blackmore.

But what if you were able to be lifted above the table, so that you could see how the balls were interacting, and you could make predictions and test them, without being part of the collisions for a little while. If only we could not only rise above, but manipulate the dynamics and conduct experiments; that is, as an independent agency to interfere with the domino effect of the Big Bang?

In fact, this is what we actually do. If we were not able to do this, the knowledge that we use to construct machines, design computers and undertake medical surgery would not exist; or, at least, the knowledge would be unusable.

If we were not able to rise above and exert independent agency of our own, we would not be able to make any invention work, other than by accident. The deterministic approach suggests that our ability to intentionally manipulate nature is merely an illusion.

To illustrate this point, this view would suggest that NASA's deliberate plan to land people on the moon and the fact that NASA landed people on the moon were a coincidence. There is no causal relationship between the events. According to the deterministic view, our conscious decision to change the domino effect of the Big Bang, to create greater complexity/order out of lower complexity/order (for example, to breach the *second law of thermodynamics*[xxii]), is an illusion. Our consciousness has no agency, because—so the argument goes—it does not exist. This would also mean that our ability to rise above na-

---

xxii The second law of thermodynamics states that in a natural thermodynamic process, there is an increase in the sum of the entropies (disorder) of the participating systems.

ture, or space-time, to observe, test theories and apply the knowledge gained to different situations, is also an illusion.

Given our understanding of physics and electronics etc., and that we can use this understanding to make things that actually work (such as computers, cars, planes, smart phones, fMRI[xxiii] machines and a million other things that appear to work reasonably well), the claim that it is merely a coincidence that these things work is an extraordinary one indeed.[xxiv] It would also be self-refuting, as it would suggest that our knowledge that we have no knowledge is also an illusion.

The lunar module Eagle. A determined universe would suggest that our plan to land people on the moon and us landing people on the moon were a coincidence, as we have no way of standing apart from the action-reaction mechanics of space-time, to gather knowledge about the system independent of it, and to use that knowledge to interfere with the domino effect that was let loose at the Big Bang.

The fact that our inventions appear to work so often shows that humans do, in fact, have a capacity to 'rise above', make predictions and channel energy independent of the chain reaction of the Big Bang. This is no small phenomenon. It is truly a profound one and points to a capacity to work outside the space-time dimensions of the Big Bang. It supports the existence of our consciousness and its position both within and external to space-time. The position of consciousness outside space-time seems to enable independent agency, that is, *free will* (at least from space-time).

---

xxiii Functional magnetic resonance imaging.
xxiv As astronomer and author Carl Sagan once wrote, 'Extraordinary claims require extraordinary evidence'.

Animal Behaviour

Awareness from outside space-time is not restricted to humans. Consider, for example, whether a wildebeest is running from a lion on the Serengeti because it has a chemical impulse to run from lions (classical view) or because it is aware that if it doesn't run from the lion, it will be killed by the lion; and being killed is not something it wants for *itself*.

If one believes that the wildebeest has an awareness of self that exists beyond the action–reaction material world, and that this appreciation plays a role in directing its behaviour to survive, it would be difficult, if not impossible, for one to also hold the seemingly contradictory view that the wildebeest's self-awareness is an illusion without agency.

On closer inspection, the behaviour of animals like the wildebeest (and, indeed, humans) appears to be informed by a perspective that exists outside the domino dynamics of space-time.

Ah, you say, but it can be learnt behaviour. Right you are. But, as we have also discussed, the ability for us to possess knowledge relies on our fundamental ability to allow our mental faculties to rise above our everyday dimensions, to see existence from a different angle. No matter which way one turns, consciousness is where the path appears to lead.

Consciousness may therefore not be a phenomenon confined to the human condition, but one intimately connected to the phenomenon of life in general.

Mind over matter—the placebo effect

'Your faith has made you well'[xxv]

The view that consciousness does not exist[23] and that any notion of it is an illusion[24] encounters further difficulty in the face of another top-down effect humans are well aware of.

[xxv] Mark 5:34. Jesus Christ, the Bible.

The *placebo effect* is a well-evidenced phenomenon which describes how a person's belief that they are taking a remedy, actually makes them better (even physically) when they are slipped a fake remedy (for example, a sugar pill) instead of the real medicine. Medical science still considers the effect a mystery, but few serious researchers deny its existence.

Over the past fifty years a mountain of evidence has gathered showing the reality of the placebo effect. For example, recently drug companies were exposed for not releasing their full studies of Prozac (a popular antidepressant). Had they done so, it would have shown that, while patients who took Prozac suffered significantly less depression, the effect was no greater than in those patients who took the fake (placebo) drug.[25]

The placebo effect shows the agency of top-down causation; that is, higher order thinking can affect lower order mechanics. Source: Jesus healing the sick by Gustave Doré, 19th century.

The placebo effect has not only been observed in drug-taking. Research has shown that it can work just as effectively in surgery.[26] As far back as the 1950s, experiments have described the placebo effect in surgical operations. For example, angina pectoris is the name given to recurrent pain in the chest and left arm due to decreased blood flow to the heart. The most common treatment is surgery, the tying of the mammary artery. In one trial, medical researchers split patients requiring the surgery into two groups. One group had the surgery in the normal way; the other group went into surgery but, instead of having their mammary arteries tied, they were just sewn back up again. Both groups of patients reported the same level of relief after surgery.[27]

In a study published in the journal *Science*, using modern imaging technology, researchers found that simply believing a pain treatment was effective actually dampened pain signalling in a region of the spinal cord called the dorsal horn, suggesting a powerful biological mechanism was at work.[28]

'It is deeply rooted in very, very early areas of the central nervous system. That definitely speaks for a strong effect,' says lead researcher Falk Eippert of the University Medical Centre Hamburg-Eppendorf.[29]

Eippert and colleagues used functional magnetic resonance imaging (fMRI) to study changes in spinal cord activity. They applied painful heat to the arms of fifteen healthy men and compared the spinal cord responses when the subjects thought they had been treated with either an anaesthetic cream or a placebo. Both creams were inactive but the fMRI scans showed that nerve activity was reduced significantly when subjects believed they were getting the anaesthetic.

Australian researchers claim the 'placebo effect is a real and therapeutic psychological phenomenon that, with more research, could be exploited more systematically in medical practice'.[30]

Damien Finniss of the University of Sydney says, 'There is not one placebo effect, but many.' In the medical journal *The Lancet*[xxvi], Finniss and his colleagues demonstrate how the placebo effect can exist even in the absence of an actual placebo. 'The act of receiving a pill brings a whole "psychosocial context" into play, which includes not only the patient's desires and expectations but also those of the clinician, and their interaction with each other and their physical and social environment.'

Finniss and his colleagues cite a randomised, single-blind, three-arm trial that enrolled 262 patients with irritable bowel syndrome, published in 2008 in the *BMJ* (British Medical Journal).[31]

[xxvi] *The Lancet* is a weekly peer-reviewed general medical journal. It is one of the world's oldest and best-known general medical journals and has been described as one of the most prestigious medical journals in the world.

- Group A: No treatment.

- Groups B1 & B2: A placebo treatment (a sham acupuncture device).

With the B1 group, the clinician inserted the placebo needles into an arm as quickly and with as little interaction with the patient as possible.

With B2, the clinician inserted the needles while following a script that sought to convey attention, warmth and confidence as well as, occasionally, thoughtful silence.

Response rates in the no-treatment, placebo-alone and placebo plus supportive care groups were 28% (no treatment, A); 44% (placebo treatment, B1); and 62% (placebo plus comforting attention, -B2). The probability that the results were the results of a random distribution was $P<0.001$ (one chance in a thousand that the placebo effect did not occur).

They also noted that certain patient personality traits—such as extroversion, and openness to experiences—were associated with heightened responses when they received supportive care in addition to the placebo treatment, but not with the placebo alone. Indeed, outcomes also differed with the practitioners.

The studies also showed a significant *nocebo* effect, in which patients experience adverse results from a fake treatment after being told it will have negative effects.

Finniss and colleagues also compared the results of *open* tests (when drugs are administered overtly to the patient) and *hidden* tests (drugs administered without the patient knowing when). These trials have two arms.

One arm (A) in which a clinician injects a drug directly into the patient in a clinical setting.

In the other arm (B), a computer-driven pump delivers the drug silently, outside the normal clinical setting and without the patient's awareness (although the patient knows that the drug will be delivered at some point).

Several studies like this have shown that drugs are less effective when given by hidden methods. Finniss and colleagues explained, 'the overall outcome of a treatment combines the specific pharmacological or physiological action of the treatment and the psychosocial context in which it is delivered'. In other words, the psychological effect of the giving of the drug may be a factor accounting for the effectiveness of all drugs, placebo or not.

They noted trials of Proglumide, a drug used to treat stomach ulcers. It appeared to be more effective than a placebo against pain in a randomised trial. However, when tested in an open-hidden study, with patients who didn't know it was coming, the drug had no effect.

'Placebo mechanisms can interact with drug treatments, even if no placebo is given, since every treatment is given in a therapeutic context that has potential to activate and modulate placebo mechanisms,' the researchers concluded.[32]

The placebo effect is no small phenomenon and significantly undermines the naturalist view that consciousness is an illusion and has no agency on the body or in the world. It provides strong evidence that our conscious expectation that something is going to have a future effect, can produce a *downward causation* (top-down) capable of exerting a force which interferes with the lower-level mechanics of space-time.

Mind over Matter—Dissociative identity disorder (DID)

Dissociative identity disorder (DID), formerly known as multiple personality disorder (MPD), is a severe condition in which two or more distinct identities, or personality states, are present in—and alternately take control of—an individual. The person also experiences memory loss that is too extensive to be explained by ordinary forgetfulness.[33]

This condition has often been incorrectly labelled in popular culture as 'schizophrenia', a term which actually describes a different disorder.[xxvii] Typically, people with DID generate additional personalities,

---

[xxvii] Schizophrenia is a condition that typically involves delusions, hallucinations and confused thinking. It is not split personality.

often as a result of abuse experienced in childhood. It is thought that they generate these additional personalities as a defence mechanism, so that they can experience life through personalities that don't remember/experience the abuse.

While this is an interesting disorder in its own right, there is something even more curious about DID that baffles the medical profession. Over the past 150 years, psychiatrists have recorded cases in which physiological conditions accompany different personalities in the same person. When the body switches personalities, its physiology also changes, with accompanying illnesses etc. which can't normally be turned on and off in a body. These include the appearance of scars and welts, altered handwriting, epilepsy, changes in eye colour, allergies, colour blindness and different reactions to the same medication.

Daniel Goleman explored the research being done by psychiatrists in the late 1980s. One psychiatrist, Bennett Braun of the International Society for the Study of Multiple Personality and a psychiatrist at Rush–Presbyterian St Luke's Medical Centre in Chicago, documented many such cases.

Often one or a number of personalities of someone with DID will be that of a child. Braun describes several instances in which separate personalities in the same person/body responded differently to a given dose of the same medication. A tranquilliser, for instance, made the childish personality of one patient sleepy and relaxed, but gave adult personalities confusion and racing thoughts. An anticonvulsant prescribed for epilepsy was given to another patient but had no effect on the personalities except those under the age of 12.[34]

Another patient was allergic to orange juice in most of his personalities; only one did not break out into rashes after drinking it.

In a study published in the *The Journal of Nervous Mental Disorders*, Scott Miller of the University of Utah, a specialist on optical differences in people with multiple personalities, engaged an ophthalmologist to conduct optical tests on ten DID patients and a control group of non-patients. After each round of tests, the ophthalmologist would leave the room while the DID patient switched personalities (sometimes at will and sometimes with prompting), then return and repeat the test. The study included a control group of another ten people who, unknown to the examiner, feigned switching personalities.

The study found that there were significant changes in visual acuity, in the shape and curvature of the eye and in refraction from personality to personality in the DID patients, but hardly any among the others tested.[35]

Testing also found some clinical differences apart from the standard tests. One woman had three personalities, aged 5, 17 and 35. When the five-year-old was examined, she had a condition common in childhood known as 'lazy eye', in which one eye turns in toward the nose. The condition was not present in the 17-year-old or the 35-year-old.

Similar differences were found in other patients. 'One patient had had his left eye injured in a fight, so that it turned out,' said Dr Miller. 'But the condition only appeared in one of his personalities.'[36]

Researchers hope that learning more about how these people are able to subconsciously 'control' their illnesses might give practitioners a greater understanding of how the mind can be better used to treat physiological conditions.

Like the placebo effect, this is not good news for pharmaceutical companies but is potentially fantastic news for society as a whole, looking for innovative ways to manage the spiralling cost of health care. Many consider a paradigm change is required in our approaches to health care, particularly in Western societies, which are increasingly being seen as financially unsustainable.

Though in terms of our safari, this phenomenon serves as yet another example of top-down causation. It is an example of how the mental faculties, beliefs or attitudes of a person produce what can appear to be miraculous changes to the physiology of an individual, in a similar fashion to the placebo effect.

Mind over matter—Brain Elasticity

Research into brain elasticity also supports the existence of this top-down force, such as studies showing how people severely paralysed by stroke can retrain their brain to control robots.[37]

In his book *The Brain that Changes Itself*, psychiatrist and psychoanalyst Norman Doidge surveys many scientific studies offering convincing evidence of neuroplasticity, showing how 'our thoughts can change the material structure of our brains at a microscopic level, because the brain is constantly adapting itself. So even talking therapy or imagination can change our brains.'[38]

An example of neuroplasticity is the case of a three-year-old girl whose left brain needed to be surgically removed because of serious chronic encephalitis. Doctors at John Hopkins Hospital in Baltimore have performed this kind of operation on hundreds of children. While the results for adults would be disastrous (permanent paralysis down one side and language deficiency), a year after her operation the girl showed almost no side effects of the procedure. She is now doing well at school and living a normal life.[39]

This appears to be an example of the bigger, immaterial brain (or mind) exerting a force on its lower-order mechanics to make it do something it hadn't been programmed to do, at least in a bottom-up deterministic way. If the components drive the neuroplasticity, then the removal of the components should result in no remedy. Without a tangible consciousness, it is very difficult to imagine an explanation for how a brain can change its hardwiring to accommodate new functions.

Emergent properties pushing down

As discussed above, the emergent property of matter, coming about as a result of the combination of trillions of non-material particles in formation, creates composite structures like rocks and jets. As Hawking suggests, the Newtonian or natural laws are a good approximation for describing the way macroscopic objects made up of quantum components behave.[40]

The question then becomes, can the laws of nature affect what happens at the quantum level, even though the 'laws of nature' are different? Can a dynamic, caused in our everyday space-time, ripple into quantum mechanics?

In fact, this is very much the case. Our space-time world, even if inspired by a perspective outside space-time (for example, consciousness), seems to determine what happens at a microscopic level. When one boils water to make tea, this action makes sense only on a macro scale. The goal is not a microscopic goal, rather it is a macroscopic one.

What happens at a microscopic level (such as the excitement of the individual molecules of water in the pot) is determined by conditions in the space-time dimension: the mechanics of boiling water to make tea. In effect, this constitutes a top-down causation. The virtual world of matter *determines* what happens at the more primal level of quantum mechanics.

It is the unique qualities of the macro orientation of millions of subatomic particles that push back down on the quantum universe, making quantum particles do things that, left to their own devices, they would not. This makes a lot more sense than the suggestion that, at a subatomic level, a collection of atoms wanted to make a cup of tea.

Eugene Wigner suggests the symmetry of physics implies that the action of matter upon the mind must give rise to 'direct action of mind upon matter'.[41] One could imagine a cascading effect; that is, the body suggesting a desire for tea, the mind's decision to make a pot of tea and the space-time mechanics of making the tea, which in turn orchestrates the microscopic world. Perhaps this is the full circle. As we have discussed above, drilling down further may reveal only mental processes.

The Agency of Ideas

'Ideas are far more powerful than guns. We don't let our people have guns. Why should we let them have ideas?' —Joseph Stalin.

Another demonstration of our ability, as living organisms, to exert a top-down force can be seen in the way we bring ideas into reality.

Philosophers have played with different theories of what ideas are. Plato thought ideas had independent form. Like some quantum physicists, Plato suggested that matter is, at a primal level, more like ideas than things. Plato had a suspicion that ideas had matter-like qualities (thousands of years before quantum mechanics).[42] It is interesting how our exploration so often leads us back to where we began.

Victor Hugo also commented on the agency of ideas, 'there is nothing more powerful than an idea whose time has come'. As Sir John Polkinghorne suggests, 'indeed, it is logically difficult to deny that ideas really exist: after all the "idea" that ideas don't exist is, at least on the face of it, self-refuting'.[43]

Strange as it might seem, there is ex fperimental evidence to suggest that ideas (or at least some of the things we treat as ideas) have an existence in their own right, even though they may not be made of matter (but then again, as we have seen, nothing much is).[44]

'There is nothing more powerful than an idea whose time has come.' Victor Hugo.

Studies by Rupert Sheldrake and others have produced evidence that once an idea is grasped, it is suddenly grasped more quickly around the world by others in a similar situation. Scientists have observed this behaviour in animals (for example, rats) and in people, such as the *Flynn Effect* in IQ tests.[45] Once a puzzle is solved, others who are not connected in any normal way, suddenly start solving the puzzle.

Sheldrake has also shown this effect in matter construction, such as in crystals, through a phenomenon he calls *morphic resonance.* As Sheldrake's research suggests the possibility that memory is inherent in nature and that natural systems—such as termite colonies, or pigeons, or orchid plants, or insulin molecules— inherit a collective memory from all previous things of their kind. Sheldrake suggests that morphic resonance provides non-local quantum-like connections between organisms.[46] This means connections are not limited by time and space (distance), a kind of entanglement relationship.

Plato suggested matter was more like ideas long before quantum physics. Plato (left) and Aristotle (right), a detail of THE SCHOOL OF ATHENS, *a fresco by Raphael.*

Sir Roger Penrose suggests that mathematics provide another clue. He noted that some of the basic physical 'laws' were 'precise to an extraordinary degree, far beyond the precision of our direct sense experience of the combined calculational powers of all conscious individuals within the ken of mankind'. He provides examples of Newton's gravitational theory as applied to the movements of the solar system, which is precise to one part in ten million. Einstein's theory of relativity improved upon Newton by another factor of ten million, and it also predicted bizarre new effects such as black holes and gravitational lenses. When scientists went looking for these unexpected phenomena, they were surprised to actually find them.[47]

Penrose suggests that the amazing accuracy of the mathematics 'was not the result of a new theory being introduced only to make sense of vast amounts of new data. Rather, the extra precision was seen only *after* each theory had been produced.' Is it possible that pure mathematics is in contact with a Platonic[xxviii] realm of ideas and forms, suggesting they're interdependent?[48] What a strange existence!

Let's think about it for a moment. Ideas do arrive in people's minds and shape behaviour.[49] Most of us at least accept that our experience of an idea is real. In fact, the philosopher and mathematician René Descartes maintains that our ability to experience an idea proves our existence. In the seventeenth century he famously described that he could doubt everything around him but the fact that he doubts; thus

---

xxviii Plato's ideas of the forms of ideas.

he (who doubts) must exist. He coined the phrase 'I think, therefore I am'.

How does a brain locked into a space-time domino effect generate innovative ideas? What is imagination? How does the brain recognise an idea and realise its value? Does the brain use the evolutionary approach and just produce lots and lots of mutations until it stumbles upon the right one, or is there something more deliberate about it? These questions are difficult to answer without a macro-phenomena such as our consciousness, which could be an interface between our brain and ideas.

Penrose and Stuart Hameroff, an anaesthesiologist and professor at the University of Arizona, are both engaged in consciousness research. Together they suggest that tiny brain structures known as *microtubules* are small enough to apply to quantum mechanics for computation.[50] They maintain that this could prove to be the brain's interface with a quantum reality, a platform for consciousness, which is not dependent on the physical brain for existence and which could account for certain exceptional phenomenon like near-death experiences (NDEs), and the survival of consciousness after death, as will be discussed later.

Does our consciousness use quantum mechanics, as Penrose and Hameroff suggest? Does it collapse all possible ideas in 'superposition' about a subject into the one that best fits? Is it a quantum computer?

Perhaps it is our consciousness that identifies an idea's value and then decides to bring the idea into reality by, for example, speaking it, or doing it, or writing it down. To do this, the conscious mind must create a downward causation, a downward force which directs or orchestrates millions of individual chain reactions within our brain to direct our bodies into action (within space-time dimensions). These forces may be small but, if applied consistently across billions of delicate chemical and electrical reactions, it isn't difficult to image the significance of the effect.

To illustrate, consider the effect of wind on water. A slight breeze across an ocean rock pool might have a negligible effect. However, spreading a slight breeze over hundreds of kilometres and billions of water molecules will create waves that can pound the shores and erode rock.

'In every block of marble I see a statue as plain as though it stood before me...' 1475. Portrait of Michelangelo at the Time of the Sistine Chapel. c.1504–06.

Michelangelo's Pietà, St Peter's Basilica (1498–99).

In this way a downward force might create a breeze-like effect over the eight billion or so neurons in an average human brain, creating an energy wave that influences the firing of neurons, just as a breeze might interfere with the molecules of sea water over hundreds of kilometres.

It is peculiar how we can so readily believe that a magnet or electron or, indeed, any material object (through gravity) can possess an invisible field of force around it, while at the same time find it so difficult to contemplate that something so complex and powerful as our own minds might have such a field as well—outside of Hollywood movies, of course. Indeed, we still don't understand *fundamentally* what magnetism is,[51] yet we accept it without question.

We see in our everyday experiences how our higher-order functions of imagination and inspiration can direct lower-level mechanics. As Michelangelo said about his sculpturing, 'in every block of marble I see a statue as plain as though it stood before me, shaped and perfect in attitude and action. I have only to hew away the rough walls that imprison the lovely apparition to reveal it to the other eyes as mine see it' (1475).

It is difficult to imagine that Michelangelo was engaged in a bottom-up task, exercising his genetic program to undertake the hewing, one stone chip at a time, as a reaction to environmental stimuli.

It makes more sense to think that a mental image or idea of the finished statue *preceded* the hewing, and gave it meaning. Our ability to capture such ideas and bring them into our space-time dimensions is probably another reason why Einstein thought that imagination and creativity were so important to the process of education.

The agency of ideas is not restricted to human life. Animal research shows solid evidence that animals work towards attaining a mental goal. In the case of the wildebeest it can be its survival (running from the lion to stay alive); in the case of a wasp it can be the building of its cone nest; in the case of a spider it can be the spinning of its web.

Researchers have observed that when they interfere with an animal's goal attainment, the animal adapts, employing different strategies to accomplish the goal. One study tested the ability of wasps to counteract every interference by experimenters to ensure that the final construction of their nest was pure to its design.[52]

The wasps' compensatory strategies were shown to be too complex to be pre-programmed genetic reactions, because the animal or insect would need to have a genetic program for an infinite number of possible situations or interferences.

Does brain size matter?

As big and complex as a human brain may be, it is difficult to believe that it, let alone a wasp's brain, could house such a mass of information related to how to react to every conceivable environmental stimuli.

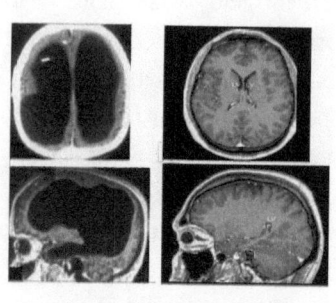

A 44-year-old man, married with kids and holding a steady job, was found to have practically no brain matter in his skull. The large black space shows the fluid that replaced much of the patient's brain (left). For comparison, the images (right) show a typical brain without any abnormalities (Images: Feuillet et al. THE LANCET *(vol 370, p 262).*

Indeed, there is other evidence to suggest that the brain is not the sole repository of our mind's information. Neurologist John Lorber, who was a specialist in the treatment of the condition hydrocephalus, found that often, despite having only a small fraction of the mass of a normal brain, hydrocephalus sufferers were not cognitively impaired.

He found several individuals who had only five per cent of the brain mass that they should have had, who seemed to have no symptoms of cognitive impairment at all. He cites four individuals who fit this description, including one with an honours degree in mathematics (who actually had substantially less than five per cent brain mass).[53 & 54]

What does this tell us about our brain's role in processing ideas and storing information? It suggests that our view of our brain as a bucket, in which we put and store information, might not be the correct view. Perhaps it is not the size of our brain which separates us from other animals—at least, not by itself.

For many, findings like those above suggest that the material collection of neurons in our brain does not fully represent the mind. After all, neurons themselves are mostly fat and ultimately made up of empty space and those strange energy particles that the rest of matter is made up of—those particles that pop in and out of our space-time dimensions.[xxix] Could Penrose and others be right in suggesting that the brain may be an interface between our consciousness and our body?

The conscious gatekeeper

From the 1980s Benjamin Libet, a research scientist at the University of California, provided convincing empirical evidence, through the use of the electroencephalogram (EEG), that the brain sends a message to move a body part (for example, a finger) 550 milliseconds before the action is carried out (what Libet termed the *Readiness Potential*).[55]

About 350 milliseconds before the action the impulse becomes conscious, and Libet's research has shown that the conscious mind has a window of opportunity of about 100 to 150 milliseconds to veto the action (primarily because the final 50 milliseconds prior to an act are occupied with the activation of the spinal motor neurones by the primary motor cortex— but you don't need to know this).

The findings show that when the subconscious mind instigates an act, the conscious mind still has a role in allowing it to occur, or not. Here consciousness appears to be acting like the final sign-off on a purchase order. Not only is it able to interfere in the Big Bang's chain reaction, but—at least in the body—the Big Bang appears to need the permission of consciousness to complete an act.

Plugging Holes in Evolution and Deoxyribonucleic Acid (DNA) Explanations

Biologists Charles Darwin and Alfred Wallace introduced the theory of evolution, including *natural selection*. Darwin launched it formally in his 1859 book *On the Origin of Species: By Means of Natural Selection*. Evolution theory describes— through chance mutation, survival of the 'fittest' and natural selection—the process which produces the variety of life on the planet.

---

xxix Of course, the question could always be posed, where do these little particles go when they disappear from our view, and why do they come back?

A portrait of 31-year-old Charles Darwin by George Richmond in 1840.

When Darwin and Wallace championed their unconventional theory, they were vigorously attacked by some of the most prominent scientists of the time.[56] Some even considered it to have radical, even revolutionary, connotations.

Nevertheless, it has not only survived but has become the dominant paradigm in the living sciences. Its modern version, which partners with deoxyribonucleic acid (DNA) science, is held by some to be the unifying theory of life sciences and is seldom questioned. Evolution has also attracted much political controversy, as it is often used by atheists to show how a creator is not necessary to explain life and existence as we know it.

There are, however, limits to the evolutionary-DNA perspective that are difficult to overcome without the addition of another dimension to our everyday reality. This was acknowledged by Darwin and Wallace. The latter, for example, pointed out that certain aspects of life could not be accounted for by natural selection, such as the creation of life, consciousness and the high mental functions of humans. 'These three distinct stages of progress from the inorganic world of matter and motion up to man, point clearly to an unseen universe ... to which the world of matter is altogether subordinate.'[57] And again, this was, of course, all before quantum physics.

Self-awareness

Firstly, evolution rests on the assumption that organisms *want*—or have a compulsion—to survive, to frame the survival instinct. In order to want to survive, an organism would appear to benefit from a holistic sense of self. Without such a sense there appears nothing to protect. The organism would be a bag of separate limbs and organs.

It is difficult to see how a strictly materialistic interpretation of evolution's 'survival of the fittest' principle could work in practice without some level of a conscious, high-level appreciation of a subjective holistic self, capable of experiencing a survival instinct and implementing survival tactics.

Limitations of the Mutations Strategy

A key dynamic in evolutionary theory is random mutation, which provides a species enough diversity in its population to allow at least some of its members to survive changes to the environment.

The difficulty is that this principle doesn't transplant well into explaining the behaviour of an organism in real time. This is important, as it has a direct bearing on survival. It would be impractical for an organism like a gazelle to try all possible or random combinations of action until it stumbled upon the right action to survive the pursuit of a cheetah.

As examined above, there is a requirement for some goal direction in order to shape behaviours too complex to have been pre-programmed reactions to environmental stimuli built up as a result of a DNA library of past 'mistakes'.

Most of us, for example, would find it difficult to imagine that an organism's DNA (for example, a gazelle's) instinctively drove it to jump to the left at a certain speed when its sight and smell detected a particular combination (for example, resembling a cheetah) at a certain distance away, in the middle of a plain with a rock in front of it of certain size, and a mud patch on the left a certain distance away etc.

A gazelle hasn't the luxury of trying all possible combinations of escape routes from a cheetah to stumble upon the right one. Source: Adapted from a photo by Bram from Breda, Netherlands.

However, by accepting the existence of a higher capacity within the organism, able to pursue a goal of survival, or avoid the idea of death, the mechanics behind the behaviour (which includes a role for an extra dimension) starts to emerge and fills this apparent blind spot in the story of evolution.

HOW A CELL KNOWS WHICH CELL TO BECOME

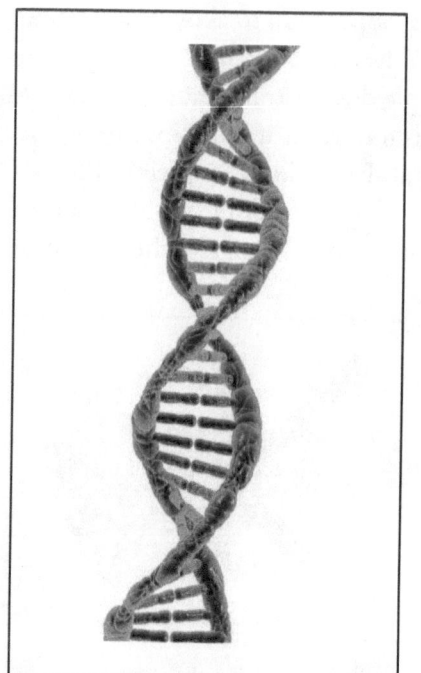

The role of an extra dimension may also help us get behind the difficulties DNA research has been having in identifying how a cell knows what to become. While research has been able to reveal the DNA instructions that provide information on the chemical compositions of living matter, there is a lack of evidence to show that DNA has any contribution to determining whether a cell becomes part of an arm instead of a leg.[58]

DNA plays a key role in the function and formation of all cells. We each have our own specific DNA and it is the only thing that is permanently with us from birth to death. From the embryonic stage, at the fourth division of cells after conception, the cell starts to specialise into cardiac, neuron, skin cells etc. The DNA is the same, but the function moves from generic to specialised. So far it is a mystery how and why this change in DNA function takes place. How can one explain the transformation of a larval caterpillar into a pupa and eventually a butterfly when its DNA structure stays the same? How can a tree be cut down and then grow again from a bark cell?[59]

A growing number of scientists, including the developmental biologist Brian Goodwin, suggest the specialisation of cell functions during the embryonic phase cannot be accounted for solely by the genetic code contained in DNA sequences. He proposes a self-organising field around the cells that can account for the differentiation and co-ordination of cells.[60]

Nobel laureate and molecule biologist Joshua Lederberg agrees that the functional differences are certainly not always a consequence of the hereditary structure of DNA but result from environmental factors around the DNA. *Epigenetics* studies reversible changes in gene function brought about without altering the DNA sequence in the cell nucleus. Research suggests that while identical twins have the same DNA, their epigenetic material can vary. Evidence is suggesting that DNA function can be determined by information outside of DNA.

This and other studies, for example, the role of DNA in determining illness and disease, have caused some to think the promises of DNA mechanics have largely been overblown.[61]

However, along with Goodwin, other biologists are increasingly considering the possibility of the existence of a kind of mould or field, invisible to the human eye, which guides the construction of cells into the constituent parts of an animal. As discussed above, Sheldrake shows evidence to support *morphic resonance*, his theory of a kind of memory mould or field, to explain this phenomenon[62]. Morphic resonance takes forward the work of several eminent scientists in the late nineteenth and early twentieth centuries, the German biologist Hans Spemann, Russian scientist Alexander Gurwitsch and Austrian-American biologist Paul Weiss. They proposed a *morphogenetic field* which helps account for the development of an embryo.[63]

The *stem cell* illustrates this dynamic well. These generic cells can grow into any cell: a heart cell, a muscle cell or even a neuron, depending on the environment. The environment appears to contain information which the cell needs to develop and specialise. When stem cells are among heart cells they will develop only into heart cells. Is there a field beyond the detection of our instruments that directs the stem cell on what to become? If so, what is the nature of such a field? What is its role in communication?

Pim van Lommel, an internationally respected cardiologist,[64] explains that all cells communicate with each other and with consciousness fields via resonance, electromagnetic fields, hormones and messenger proteins. Studying cell communication reveals speeds of communication that are difficult to understand only through electrochemical reactions.

Cell collaboration within the human body that enables continuity of function of various cell systems must occur within nanoseconds, given the replacement of 500,000 cells each second. Due to the distance between the various cell systems in the body, the speed of this information exchange must approach the speed of light, which is much faster than chemical reactions alone could manage.[65]

Signs of the phenomenon have also been recorded in other areas of biological research; for example, showing that pieces of DNA can recognise matching DNA strands at a distance without direct physical contact.[66]

Goodwin and others say this phenomenon can account for the huge number of reactions occurring in living organisms a million times faster than they occur in the laboratory. Field communication and perhaps *entanglement*[xxx] (as a result of all cells coming from the one source) may account for high-speed communication between cells but also in superorganisms; that is, organisms which in large numbers start acting like a single entity, such as bees, ants and termites.

When a queen is isolated from the colony but alive, everything continues as normal. But if the queen is killed away from her colony, chaos ensues and work stops. The queen appears able to coordinate *non-locally*[xxxi] through a collective consciousness seemingly facilitated by a shared DNA.[67]

This phenomenon can also be observed in flocks of birds and schools of fish. A flock of birds recorded on film was found to have a

---

[xxx] Recall *'Entanglement'* in quantum physics where two entangled objects can act as though they are physically connected even though they are separated by vast distances.

[xxxi] Recall *non-local*, the term that describes in physics the ability to behave as though there are no time or space barriers; that is, a realm in which space and time don't count.

reaction speed of 38 milliseconds, which is far too quick for normal communication between hundreds and thousands of birds often many metres apart.[68]

This kind of phenomenon is making quantum biology one of the fastest expanding fields in biology.

'There are definitely three areas that have turned out to be manifestly quantum,' says Dr Luca Turin, of the Fleming Biomedical Research Centre in Greece. 'These three things ... have dispelled the idea that quantum mechanics had nothing to say about biology.'[69]

The navigational feat of birds that cross countries, continents or even fly pole to pole is a good example.

European robins orient themselves for migration using specific wavelengths of light, and very weak radio waves can completely mix up their sense of direction. Yet these factors should not affect the standard compass that biologists used to think birds had within their cells.

The way birds sense direction could be something like a heads-up display in their eyes. They see magnetic fields. Three trumpeter swans flying at Riverlands Migratory Bird Sanctuary, Missouri, USA. Credit: Snowmanradio.

What makes more sense is the quantum effect of entanglement. Under quantum rules, no matter how far apart an 'entangled' pair of particles gets, each particle seems to *know* what the other is up to—they can even seem to pass information to one another faster than the speed of light.

Experiments suggest this is going on within single molecules in birds' eyes, and John Morton, of University College London, explains that the way birds sense could be stranger still. 'You could think about that as ... a kind of "heads-up display" like what pilots have: an image of the magnetic field ... imprinted on top of the image that they see around them,' says Morton.[70]

The most established of Turin's three areas is, however, photosynthesis—the staggeringly efficient process by which plants and some bacteria build the molecules they need, using energy from sunlight. It appears to use what is called *superposition* (seeming to be in more than one place at one time).

Little packets of energy simultaneously 'try' all of the possible paths to get where they need to go, and then settle on the most efficient, much like the wave form of a photon, before becoming a material particle.

'Biology seems to have been able to use these subtle effects in a warm, wet environment and still maintain the "superposition". How it does that we don't understand,' says Richard Cogdell, of the University of Glasgow.[71]

There is also the suspicion in science that quantum effects are taking place in our nose. Most smell researchers think the way that we smell has to do only with the shapes of odour molecules matching those of receptors in our noses.

But Turin believes that the way smell molecules wiggle and vibrate is responsible—thanks to the quantum effect called *tunnelling*. It suggests electrons in the receptors in our noses disappear on one side of a smell molecule and reappear on the other, leaving a little bit of energy behind in the process.

A paper recently published in *Plos One* shows that people can tell the difference between two molecules of identical shape but with different vibrations, suggesting that shape is not the only thing at work.

Lommel references tests done by the inventor of the polygraph (lie detector), Cleve Backster, which shows how white blood cells transported twenty kilometres from their owner, or when placed in locations shielded from electromagnetic radiation (in a Faraday cage), were found to react when the subject was shown horrific or sexually arousing images. Anomalies in the subject's skin conductivity were found at the same time. He offered proof of non-local communication between consciousness and white blood cells.

There is also the phenomenon of organ transplant patients sensing snippets of feelings and ideas that turn out to match the personality and consciousness of the deceased person, as researched for the 2003

Discovery Channel documentary called *Transplanting Memories*, by Claire Sylvia in her book *Change of Heart*, and by physician Paul Pearsall in his book *The Heart's Code*.[72]

Other researchers have claimed that they have been able to photograph electromagnetic residue of such a field, for example, through *Kirlian photography*. This uses high voltages and photographic plates to capture the image of a body or plant's electromagnetic signature, or corona. When parts of the plant are cut off, the phantom images linger for a time. However, others claim that when put to the test, the effect is merely the result of water residue, rather than evidence of an aura.[73]

Down to earth about DNA

Despite the intricacies of advanced science, perhaps the everyday examples provide the best comparisons: on the basis of DNA, chimpanzees are 96–98 per cent the same as humans[74]. Yet, despite this, most of us recognise that the real difference is likely to be greater.

I concede, however, that we do look more like chimpanzees than, say, mice. The problem is that as far as DNA goes, 99 per cent of the genome of mice is the same as that in humans.[75] So we can see the need to dig deeper for our answers, as there appears to be a distance between a solely DNA evolutionary explanation for life and the complete explanation, whatever that may be.

You might recall above, that the discovery of dark matter and energy suggests we really only know (or experience) about four per cent of the universe, the matter we are familiar with. Which means we need to be humble, particularly about physics. But did you also know, by coincidence, that this is exactly the same amount we know about DNA? Ninety-six per cent of DNA has no known biological function, earning it the label 'junk DNA'. Again this suggests we need to be humble about our understanding of biology, as the fact is that the more science discovers about the nature of reality, the more we discover how little we actually understand.[76]

Values

Another difficulty evolution has is the explanation of values (in the sense that we know them) as something that rises above practicality. Without an additional dimension, able to add a layer of higher-order meaning, there are no acts of genuine kindness (altruism);[77] no one would go to war out of a sense of duty, loyalty, justice or even self-preservation; cruelty to children would in essence be no different to having a haircut; and our legal system could not find anyone guilty of an offence (any self-respecting judge would have to blame everything on the Big Bang).

Artist's sketch of Daniel M'Naghten and an engraving of his signature. 1843.

The legal standard used in many countries for the defence on grounds of insanity was laid down as a result of the case of Daniel M'Naghten. A jury found M'Naghten not guilty of the shooting and killing of the British Prime Minister Robert Peel's secretary in 1843 by reason of insanity. He was found by psychiatrists to be under the delusion that the Prime Minister and the Pope were conspiring against him. He travelled to London to attack the Prime Minister but accidentally killed the secretary. John Martin of the *Washington Post* writes, 'The M'Naghten rule says defendants may be acquitted only if they laboured "under such defect of reason from disease of the mind" as to not realise what they were doing or why it was a crime. Some call it the "right–wrong" test.'[78]

The acceptance of the insanity plea in our legal system seems to support the concept that there is an institutional acceptance of the existence of a higher-order mind, or 'decision-maker', which has the ability to make choices above the determined space-time dimension—for example, of right and wrong. It acknowledges that parts of the mind, or perhaps brain, have a role in providing the decision-maker with information it needs to make a choice, although these mechanisms can become corrupted. In these cases the decision-maker's responsibility to make the 'right' choice is diminished. It is difficult to imagine how a deterministic view of reality would allow such a distinction between the two aspects of the mind; that is, the *provider* of information and analysis and the *decision-maker*.

The courts also acknowledge that there is a right and a wrong, by which action can be judged, over and above practicality. This is not merely a theoretical or even mathematical assessment but originates from our ability to perceive a higher level of meaning. Most of us who do not score high on the psychopathy scale can sense something fundamentally different between cruelty to children and practical things like taking the shortest route home.

In our courts, the assumption is usually made that a person has an intuitive understanding of right and wrong. Rarely do we see cases investigate what a person was taught at school or university or by parents about, for example, the merits of not stealing or murdering. Except in extreme cases, such as insanity pleas, it is usually assumed that because of their humanity, a person is able to appreciate high-order values. They are able to understand that murder, stealing etc ... are wrong and that they were able to make a conscious decisions to override normal genetic or life differences if necessary, including evolutionary survival-of-the-fittest impulses, to make the *right* decision. If they do not do this, then a punishment is administered. Evolutionary DNA 'survival of the fittest' pleas are not a serious defence in a court of law. The suggestion that a person lacks the faculties to appreciate higher-order meaning would imply to a civilised court that the person was insane.

Some modern evolutionists have suggested it possible to get around the existence of higher-order principles or values—like justice etc.—by suggesting that evolution might have produced small physical gene-like cultural things[xxxii] in our brains, which propel our actions in social dynamics; though no one has ever found any physical evidence of these things. It is proposed also that altruism can be explained by misfiring in the brain. Biologist Richard Dawkins, for example, calls them 'blessed mistakes',[79] a curious description to come from such a fierce atheist.

Again, if there were another dimensional reality able to apply meaning to reality—that is, a 'value' to a situation or action in relation to attaining a future (even virtual) goal—then the human experience of values, as something beyond the seemingly practical, has a possible explanation (more on Values later).

INTELLIGENT DESIGN

There is also the debate between atheist evolutionists and religious creationists around *intelligent design* and *irreducible complexity*. Biochemistry professor Michael Behe coined the term 'irreducible complexity', which he defines as a complex system 'composed of several well-matched, interacting parts that contribute to the basic function, wherein the removal of any one of the parts causes the system to effectively cease functioning'.[80]

Though criticised by evolution theorists and frowned upon by a large part of the scientific establishment,[81] proponents of irreducible complexity argue that it is difficult to explain complex organs, such as the human eye,[xxxiii] by relying only on the accumulation of chance mutations.[82]

The argument suggests that so many seemingly useless parts need to come together before the mechanism has any use, and that the parts

---

xxxii For example, Dawkins's 'memes'.

xxxiii Darwin wrote in *On the Origin of Species*, for example, 'to suppose that the eye ... could have been formed by natural selection, seems, I freely confess, absurd in the highest possible degree' and could only be possible if its gradual evolution could be shown, which later in his book he tries to imagine. The difficulty is that our fossil history does not appear to support *gradual* evolutionary changes in the development of species.

would not survive the evolutionary process to get to the final stage. Evolution on its own can't explain the coming together of some things.

It may remind some of William Paley's 1802 watchmaker example. Paley pointed out that if while walking across a field you came across a watch, you might think that there must have been a watchmaker, as the possibility of a watch being assembled accidentally by nature would seem ridiculously minuscule.

The argument suggests that there are organisms in existence so complex that their existence could be equated to the discovery of Shakespeare's play *Hamlet* carved in a giant stone on Mars. It might be theoretically possible that chance winds, sand and rain etc. might have come together accidentally over thousands of years to carve out an English text of the play on Mars, but most people would find the suggestion a stretch in the extreme, even if we do humour the multiverse theory (as there are an infinite number of other universes, something like this is, like us, quite possible).

Evolution theorists present arguments and research to explain how it is theoretically possible for complex organisms to arise out of evolutionary processes,[83] with some of the more enthusiastic suggesting either the multiverse or the estimated number of planets in our universe as an explanation.[84] Some suggest that after the universe, life and perhaps the eukaryotic cell (human kind of cell) are created, chance operates differently for evolution.

There are also some who object to intelligent design in principle, fearing it might make science lazy (that is, whenever we come across something difficult to explain we just call this 'God's creation' and move on).[85]

It is an argument similar to those used against atheists believing in the primacy of evolution; for example, the danger that such beliefs would make genocide or the sterilisation of handicapped people permissible under the slogan 'survival of the fittest'. 'The fittest' is often code for 'the most powerful', or even 'the supreme race'. But as Wallace was quick to point out, 'those who succeed in the race for wealth are by no means the best or most intelligent'.

While an in-principle rejection of evidence seems unscientific, there are occasions when it is demanded. Many consider that the research of Josef Mengele, an officer of Hitler's Schutzstaffel (SS) and a physician in Auschwitz concentration camp during World War II, represents evidence that falls into such a category. Mengele selected victims to be killed in the gas chambers and performed unscientific and often deadly experiments on them. Today there is a widely held view that, since Mengele exposed these victims to such unconscionable treatment, the lessons learnt through his method should not be included in medical knowledge.

This type of in-principle rejection is understandable, but to say that it *might* make science lazy would be to take things a little too far. It also seems to ignore the fact that science was developed out of religion (more later).[xxxiv] However, it is something to keep an eye on nevertheless if intelligent design becomes more established.

It is odd, though, that some species or organisms are presented as explainable while others are not. The larger challenge would seem to lie in explaining even basic life forms or organisms, even the creation of the universe before life began. The suggestion that these are somehow explainable while others are not overlooks the monumental gaps that already exist in contemporary biology and cosmology, as discussed above.

COURT CASES ABOUT WHAT IS TAUGHT IN US SCHOOLS

As with many of society's issues, school curriculums can become the battleground for politics. In the United States a number of court cases have considered whether the teaching of intelligent design represents a violation of the constitutional First Amendment (separation of church

---

[xxxiv] As will be discussed later, science, and particularly the study of nature, was in many respects conducted because of, rather than in spite of, religious beliefs, as reflected by some of history's greatest scientists, such as Newton, Einstein, Galileo, Darwin and Planck, just to name a few. As will be discussed later, Islam was first to discover the scientific method.

and state) due to its potential to support creationism, that is, the belief in a God who created our existence.

Ironically, leading and supporting court cases against the introduction of intelligent design in schools were some high-profile atheists determined to ensure that evolution was the only view presented in the classroom. They applauded the judgement in a particular case in 2005 when a US District Court judge, John E. Jones III, stated in *Kitzmiller v. Dover Area School District*, that 'we have concluded that Intelligent Design is not science, and moreover that ID cannot uncouple itself from its creationist, and thus religious, antecedents'.

It was a curious decision. The irony increases when one considers, as we shall see later, that science was created by religion and evolution sprouted out of a naturalist theme developed by scholars such as Sir Isaac Newton, whose scientific pursuits were inspired by his profound belief in a rational God. Indeed, history suggests that it is likely Newton considered himself more theologian than physicist.

As discussed above and will be later, one of the co-founders of evolution, Charles Darwin, was a believer himself (deist[xxxv]) and took exception to his theories being used as a weapon against religious belief. In fact the other founder, Alfred Wallace, concluded in *Darwinism*, that, 'We thus find that the Darwinian theory, even when carried out to its extreme logical conclusion, not only does not oppose, but lends a decided support to, a belief in the spiritual nature of man.'[86] So one might be forgiven for thinking that there are arguably more grounds for asking atheism to separate from evolution than asking science to separate from 'religious antecedents'.

Since the above case, US law and court cases have ensued on the topic, for example, in Tennessee, when in 2012, a bill was passed protecting 'teachers who explore the "scientific strengths and scientific weaknesses" of evolution'. Generally, there now appears to be more weight placed on providing children a mixture of views. Which is probably a good thing.

---

xxxv The belief that God has created the universe but remains apart from it and permits his/her creation to administer itself through natural laws.

Room for subspace

There are other challenges for evolution theory. Rational choice theory (that people act rationally in self-interest) has framed social and economic modelling for some time. Its principles have flowed over to biological modelling, including evolutionary theory and its extensions, such as theory around rationally self-interested genes.[87]

As discussed in a previous section, research has now shown rational choice theory to be seriously flawed. This has naturally undermined other theories which it has been used to support, including *modern* evolution theory, which benefits from the idea that organisms make rational choices out of self-interest.

Evolution theory also has difficulty making predictions[88] and lacks the ability to be disproved[xxxvi] (which causes some to consider it more like a philosophy or religion than a scientific theory). But at the end of the day, the real issue is that nobody knows what actually happened in the soup from which life sprouted. So until we do know, the debate will continue, as it does for those many other unanswerable questions, like how the Big Bang came about, and how a bicycle works.[106] [xxxvii]

---

xxxvi 'How do you know evolution created a toe? Because a toe exists. How would a toe look if evolution did not create it? Why has rhino eyesight evolved to be so poor? Because it helps them survive. How do we know it helps them survive? Because evolution has selected them for survival, with bad eyesight. How would rhino eyesight be if evolution did not create it?'

xxxvii 'Forget mysterious dark matter and the inexplicable accelerating expansion of the universe; the bicycle represents a far more embarrassing hole in the ac-

Notwithstanding the elusiveness of definitive answers, the evolution of life is not above breaching space-time limits, from quantum behaviour in living mechanics to conscious goal attainment. Contrary to consciousness being a foe to evolution, a more intelligent approach would consider the complementary nature of consciousness/subspace to evolutionary/space-time mechanics. It may provide some insight into how the universe can, through us, be alive and able to perceive itself (after all, we are part of the universe). As Wallace writes, those that accept his and Darwin's evidence 'will be able to accept the spiritual nature of man, as not in any way inconsistent with the theory of evolution, but as dependent on those fundamental laws and causes which furnish the very materials for evolution to work with'.

Not confined by the skull

In support of the view that the mind is not restricted by neurons and chemical reactions, there is convincing evidence that our consciousness may not be confined to our skulls[89] and may channel itself through another subspace-type dimension to affect different locations outside of our bodies.

---

complishments of physics.' *http://www.newstatesman.com/2013/07/mysteries-bicycle*

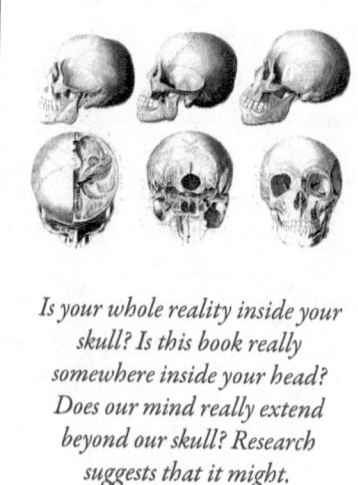

Is your whole reality inside your skull? Is this book really somewhere inside your head? Does our mind really extend beyond our skull? Research suggests that it might.

Research has repeatedly shown that people have a capacity to perceive they're being stared at. Dean Radin, a parapsychology researcher who conducted classified research for the US Government while at SRI International, undertook a meta-analysis of research conducted by Rupert Sheldrake and others testing people's ability to tell whether they were being watched. It showed an effect that would only be attributed to chance at a staggering probability of 202 octodecillion to 1, (that is, $2 \ 10^{57}$ to one). Take away 53 zeros and you still have an astonishing result.

When you compare this with the probability demanded in medical trials—100 to one—the research of Sheldrake et al. would appear quite conclusive.

Indeed, various forms of martial arts teach people how to improve their skills in doing this; people whose profession involves the observation of other people have provided testimony to the phenomenon, from soldiers to security staff.[90]

Again, without a consciousness able to break space-time restraints through a subspace-type medium, these results would appear impossible.

Global Consciousness Project

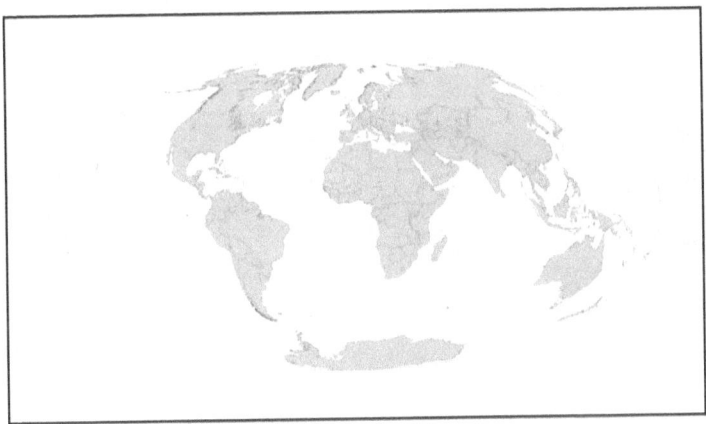

Scientists are also observing phenomena suggesting that our individual consciousness is linked in some way to the consciousness of everyone else. Testing has shown that our combined consciousness appears to have an effect beyond a space-time explanation.

The Global Consciousness Project (GCP),[91] dubbed by Roger Nelson, measures global mental coherence as a result of major global events. In short, it has found that random-number generators (computers designed to produce random numbers) suddenly stop being random at times of major global events, such as 9/11, the change of the millennium (2000), major tsunamis and the death of Princess Diana.

The data shows that when the world's attention comes together it has a physical effect on the world, as detected by random computer generators (continuously running at various locations around the world as part of the project). Cumulative analysis of the tests from 1998 to 2005 confirm a one-in-36,400 chance that the effects are random.[92] Overall Global Consciousness Project results now suggest a 1 in 3 trillion chance of being random.[93] Again, remember the one-in-100 significance required for medical trials!

Perhaps even more startling is the unexpected effect recorded—that the influence on random-number generators starts to occur *prior* to the event taking place. For example, on 11 September 2001, the effect was at its strongest two hours *before* a hijacked jet crashed into the World Trade Tower 1 in New York City at 8:46 am, and had weakened to its lowest level eight hours later.

Again, this suggests that, at some level, consciousness is not totally contained by our skulls or the 'laws' of our everyday space-time dimensions. It provides tantalising evidence that our consciousness may have access to a subspace-type dimension.

Random-number generators around the world started to behave non-randomly two hours before the jet crashed into World Trade Tower 1.

Above: north face, south tower, after plane strike (Robert on Flickr). The death of Diana, Princess of Wales, had similar effects.

Forensic evidence

There are other tell-tale signs of consciousness being outside our everyday space-time dimensions. Benford's law states that in lists of numbers from many real-life sources of data, the leading digit is distributed in a specific, non-uniform way. For example, the first digit is 1 in about 30 per cent of occurrences, and larger digits occur as the leading digit at lower and lower frequencies, to the point where 9 as a first digit occurs less than five per cent of the time.

This result has been found to apply to a wide variety of data sets, including electricity bills, street addresses, stock prices, population numbers, death rates, lengths of rivers, physical and mathematical constants and processes described by 'power laws' (which are very common in nature).

Auditors and accountants use this to detect possible fraud, because, when data sets such as those above do not show this natural distribution, it is widely recognised as a possible sign of human interference, such as fraud. The suggestion that humans are part of the natural flow of a determined universe, with no agency outside of it, appears inconsistent with this evidence.[94]

In fact, experienced investigators and those who have researched human indicators of deception will testify to the stress a body exhibits when telling a lie.[95] Telling a lie appears to be counter to the body's natural flow, and supports the existence of an agency beyond the physical body which can have an effect on the physical world, or at least interfere with its natural domino flow.

PSI

Arguably, some of the most interesting stories in science relate to researchers who have dared to investigate phenomena considered by mainstream science as 'pseudo-science' or 'non-science'. Regardless of how many Nobel prizes are given out to ideas that were once considered unscientific, challenging the prejudices of an establishment is not a task scientists take on lightly; it is known to have career-limiting properties.

Psi researchers are in this category today. Suffering all the prejudice quantum research did in the first half of the twentieth century, researchers from a range of backgrounds, from those working in classified defence laboratories to Nobel Prize winners, use the scientific method to present evidence of phenomena largely unexplainable within a Western philosophical view of reality.

Psychic (or 'psi') phenomena fall into two general categories. The first is perception of objects or events beyond the range of the ordinary senses. The second is mentally causing action at a distance.[96]

Over the past hundred years or so psi has generally been classified as phenomena which cannot be explained by our everyday senses but result in:

- information passing between different people (*telepathy*),

- a person's intention causing changes in the material world (*psychokinesis*),

- information from the material world getting into a person's mind (*clairvoyance*), or

- information entering a person's mind which has slipped in time from either the future or the past (*precognition* or *retro-cognition*).

The demonstration of such effects would clearly provide support not only for consciousness, but for a subspace-like medium traversed by consciousness to produce such phenomenon.

Probability

One method used to gauge the significance of the psi effect is to compare the results of psi testing with what would have occurred if chance alone determined the outcome, an approach used throughout science. For example, as suggested above, to show effects in medical trials, researchers often look for a probability of chance accounting for the results as 1 in 100. In psychological testing, a significant result would start at 1 in 20. This is what is referred to as 'statistically significant'.

In other areas, different probabilities are required. For example, in its Fourth Assessment Report, the Intergovernmental Panel on Climate Change (IPCC), suggested that the probability that humans are not the cause of global warming over the past 250 years is about 1 in 10.[97]

Effects

Zener cards used in the early twentieth century for experimental research into ESP.

Notwithstanding the challenges of psi research, which often deals with subtle effects,[xxxviii] the results have been astonishing, consistent and repeated. Those who doubt have only to look at the evidence.[98] Following is a list of some of the typical results you will find.

Telepathy

Tests of telepathy, that is, one person connected to the mind of another, have been done repeatedly in the laboratory over the past hundred years. One test involving one person perceiving what cards another was looking at was conducted at Duke University and showed a probability that the results were caused by chance as 1 in 10^{27}, that is, 1,000,000,000,000,000,000,000,000,000 to 1.[99]

If this surprises you, or challenges your ideas about reality, then hold on to your seat. We've only just started.

[xxxviii] People aren't predicting the lotto numbers as yet.

The Ganzfeld test

In the Ganzfeld test a person is placed into an electromagnetically shielded room under a red light. Two half ping-pong balls are placed over the eyes and the person is asked to keep the eyes open. The person sees a uniform red field and hears white noise through headphones.

After five to ten minutes people begin to hallucinate. This is because the brain becomes starved of patterns or information. As the mind has difficulty coping with a void, it begins to manufacture stuff. This person is the *receiver* of the telepathic message.

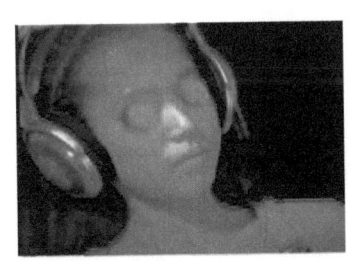

Example of a subject in a Ganzfeld experiment. This experiment has been reproduced around the world by different scientists but with the same results. ESP may be real regardless of whether you want to believe in it or not. What is it that connects us to each other? What part of reality is this? Is this subspace?

Some distance away, such as on a separate floor, there is a *sender*, a person who is given some information, such as a picture of an elephant, to send to the *receiver*. These tests are done in a double-blind way: the receiver does not know what information is presented to the sender; and the testers do not know, at the time, what information is presented to the sender.

While messages are sent to the receiver he or she is requested to *mentate*—to say aloud whatever comes to mind. If, for example, the sender is shown a picture of an elephant, it is considered a promising outcome if the receiver starts talking about 'grey' and 'animals'.

After twenty minutes in the Ganzfeld state, the receiver is brought out of it and the mentations are played back to them. They are then given a group of different items, such as pictures, among which is the one that was presented to the sender—for example, a picture of an elephant and three other pictures. The probability that the receiver would pick the same picture by chance is one in four, or 25 per cent.

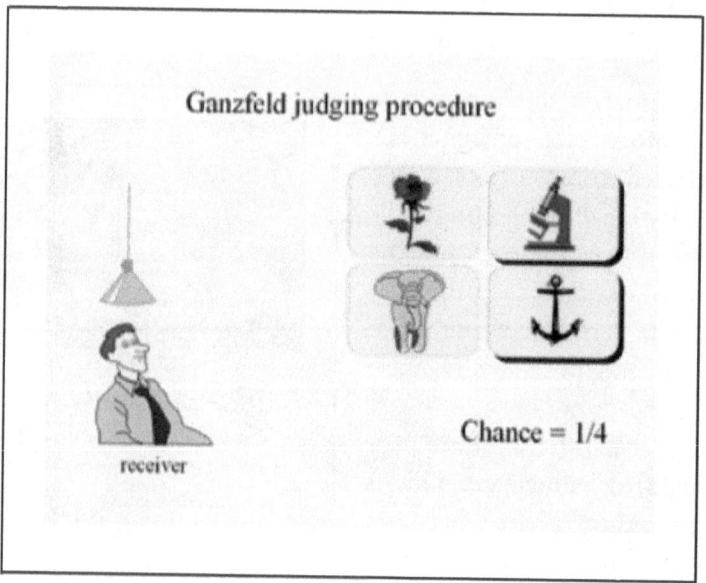

Since the 1920s this experiment has been repeated by a vast number of universities, including Harvard, Duke, Cornell, Stanford, Leningrad, Amsterdam and McGill.

Testing a sample of 122 experiments conducted at different places in the world and taking into account a total of 4,674 sessions, Dean Radin[xxxix] shows the cumulative hit rate of 32 per cent[100]. The odds against chance accounting for the results are 3×10^{-30}, that is, 300 trillion quadrillion to 1; so in order to get this result by chance, the experimenters would need to run the test three trillion quadrillion times. Translating this into time, it would take longer than the age of the universe to get these results by chance.

The experiments were also performed by scientists who professed not to believe in the effects, including Edward Degado-Romero of the University of Georgia and George Howard of the University of Notre Dame. To their apparent disappointment, they also achieved a hit rate of 32 per cent.[101]

---

xxxix Dr Dean Radin, a world renowned psi researcher who conducted classified psi research for the US Government while at SRI International.

What testing has found is that when the sender thinks of something, the presented information or picture evokes the same thought bubble in the mind of the receiver. It is as if they are somehow *entangled* (using the physics analogy).

These findings have been discussed a number of times since 1994 in the journal *Psychological Bulletin* and have shown that since the 1980s the hit rate has settled at 32 per cent with a reduced standard error, meaning that 32 per cent appears to be the actual effect.

Limiting these experiments to specific groups of participants—such as creative people or skilled meditators—has shown an escalation of the rate, for example, to around 50 per cent.[102]

Researchers have also conducted and replicated experiments tracking brain activity with EEGs and MRIs, in which the sender is shown a visual stimulus, such as a flickering light, and the brain activity of the sender and the receiver are compared.

The results show highly significant outcomes, suggesting that the receiver's brain patterns are activated in a similar—though less intense—way, even though they are not being shown the visual stimulus. Again this reminds one of the entanglement effect in physics. The results of these tests have been published in some of the most respected scientific journals, including *Science* and *Nature*.[103]

In a study carried out by Jeanne Achterberg et al. and published in 2005 in the *Journal of Alternative and Complementary Medicine*, Achterberg used functional magnetic resonance imaging (fMRI) technology to demonstrate that *distant intentionality* (DI), defined as sending thoughts at a distance, is correlated with an activation of certain brain functions in the recipients.[104]

This test recruited eleven Hawaiian traditional kahuna healers/priests who believed they could connect with or heal at a distance. Each healer chose a person with whom they felt a special connection as a recipient for DI. The recipient was placed in the MRI scanner and isolated from all forms of sensory contact with the healer.

The healers sent forms of DI that related to their own healing practices at random two-minute intervals that were unknown to the recipient. Significant differences between experimental (send) and control (no send) procedures were found (p < 0.000127)—close to a one in 8,000 chance that the results indicated no effect.

Areas activated during the experimental procedures included the anterior and middle cingulate, precuneus and frontal areas. It was concluded that instructions to a healer to make an intentional connection with a sensory-isolated person could be correlated to changes in brain function of that individual (the receiver). The experiment has been replicated a number of times since, with similar results.[105]

Remote viewing

Remote viewing (a clairvoyant ability to intuitively describe a target place or the content of a room etc. without being there or having direct experience of the place) has produced results that could only be accounted for by chance as one in 33 trillion.[106] People could describe the interior of buildings and contents of locked filing cabinets—or state secrets.

So impressive were the results of experiments carried out by physicists Hal Puthoff and Russell Targ in the 1970s, they were supported by the CIA, NASA, the Defense Intelligence Agency, and Army and Air Force Intelligence through a $20 million research program at the Stanford Research Institute (SRI) and the Science Applications International Corporation (SAIC) over 20 years. The experiments routinely presented results that had a probability of occurring by chance of less than one in a million.

In a recent interview, Targ talked about the research.

'I'm not just talking about the best ever things that happened one day in the laboratory, rather I am describing experiments that were published in the most prestigious journals of the world, *Nature*, the *Proceedings of the IEEE*, and the *Proceedings of the American Academy of Sciences*.'[107]

Targ also commented on the emergence of programs aimed at teaching remote-viewing expertise for considerable sums of money. He said, however, that remote viewing, as with many other psychic abilities, was widespread in the community:

'I'm not just talking about the best ever things that happened one day in the laboratory.' Russell Targ, 2014.

'... psychic abilities are real, they're available and they're free. You do not have to pay thousands of dollars to get in touch with your psychic abilities ... there are a lot of opportunities to pay people a lot of money to learn how to do remote viewing. The secret is that there is no secret ...'

In his book, *The Reality of ESP, A Physicist's Proof of Psychic Abilities*, Targ includes the following example of the application of these abilities:

On February 4, 1974, members of the Symbionese Liberation Army kidnapped nineteen-year-old newspaper heiress Patricia Hearst from her Berkeley, California, apartment.

Commander Deanna Troi is a main character in the science-fiction television series STAR TREK: THE NEXT GENERATION *and related TV series and films, portrayed by actress Marina Sirtis (above). Troi is half-human, half-Betazoid (a race that can read minds) and has the ability to sense emotions.*

Desperate to find her, the police called Targ and Pat Price, a psychic retired police commissioner who had demonstrated remarkable remote-viewing abilities in and out of the laboratory.

As Price turned the pages of the police mug book filled with hundreds of photos, he suddenly pointed to one of them and announced, 'That's the ringleader.' The man was Donald DeFreeze, who was indeed subsequently so identified. Price also described the type and location of the kidnap car, enabling the police to find it within minutes.[108]

Targ talks about the technique he used to show the government the effects beyond the test results. He would invite program monitors and other Washington officials into the laboratory to try the experiments themselves, and discover their own abilities. He recalled how once the head of an agency had to come to the laboratories to be tested after his staff had returned from visits and proclaimed their psychic skills.[109]

More recent examples include remote viewers describing the whereabouts of Saddam Hussein's hiding place.[110]

Beyond earthly observations, perceptions at extreme distances in space include that of a ring around the planet Jupiter that had never been seen from Earth. Its presence was confirmed only later by images taken by the passing NASA space probe Pioneer 10.

'I am convinced,' Targ says, 'that most people can learn to move from their ordinary mind to one not obstructed by conventional barriers of space and time. Who would not want to try that?'[111] As one can imagine, this provides considerable support for consciousness and its utilisation of a subspace dimension.

Psychokinesis and the quantum measurement problem

Robert G. Jahn, a professor of aerospace sciences and dean emeritus of the School of Engineering and Applied Science at Princeton University, was a consultant for NASA and initially specialised in deep-space propulsion. He is author of *Physics of Electric Propulsion*, a leading text book in the field. Jahn had no interest or belief in psychokinesis but one day reluctantly agreed to a student's request to oversee an experiment in the subject. The results were so extraordinary that he became a convert. This and other experiences inspired him to found the Princeton Engineering Anomalies Research (PEAR) lab, which has produced consistent results supporting the psychokinesis phenomenon.

An example of his early work: Jahn and his associate, clinical psychologist Brenda Dunne, used a random-event generator (REG), a device which uses radioactive decay to produce random binary numbers, to act as a kind of coin-flipper.

Jahn and Dunne arranged for volunteers to sit in front of a REG and concentrate on having the device produce more of either heads or tails. Over the course of thousands of trials they found that the volunteers were able to have a small but statistically significant effect on the REG. They also found that the ability was not restricted to a few but existed throughout the volunteers.[112]

Above we described the quantum measurement problem in physics. Basically, it is when an elementary quantum (unit) behaves like a wave when you are not looking at it and like a particle (chunk of matter) when you are.

There are different schools of thought about the relationship between our conscious mind and what is happening with the collapsing of the wave form. Some scientists believe that consciousness is not merely important but fundamental to the formation of reality, that consciousness actually collapses potentiality into reality. This is the *strong view* and has been endorsed by some of the most eminent physicists in the world, including John von Neumann, Bernard d'Espagnat,

Euan Squires and Henry Stapp.[113] It challenges the assumption that the world was here before consciousness.

Others take a 'softer' approach, including the suggestion that observation merely increases our knowledge about a measurement system. Others choose to ignore it as a 'problem' altogether, or suggest that the whole thing works once you accept we have no free will, which is a popular view in neuroscience today.

Radin decided to put this to the test. On the assumption that meditators can sustain their conscious attention for longer than ordinary people, he constructed experiments which tested the ability of meditators and non-meditators to consciously interfere with the progress of a quantum/photon.

During the tests, participants typically sat quietly, one at a time, on a chair or on the floor, about two metres from the outer wall of an electromagnetically shielded steel chamber. Each was asked to imagine that he or she could mentally 'see' the photons in the target area of the interferometer inside the chamber. Or if they found this too difficult, were asked to mentally block the photons. They were asked to do this at different intervals.

Radin used five meditators, four of whom had many decades of daily meditative practice. Other participants had either none or less than two years meditative experience.

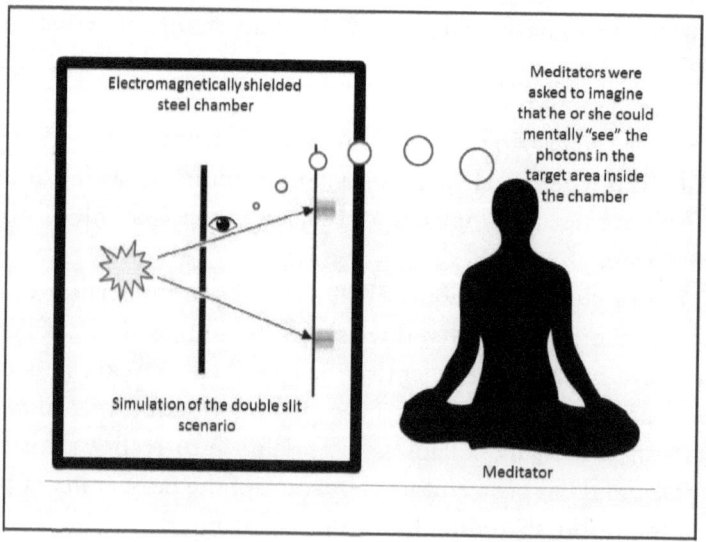

The results showed that the meditators were able to register more particle photons through the equipment at combined odds against chance of 107,000 to 1. The non-meditators' results were close to chance expectations.[114] The results backed up similar experiments conducted by others in the past, suggesting a mind-matter interaction.

To further test whether it was consciousness 'collapsing' the quantum wave function, Radin followed up the test with an actual optical double-slit system. In these experiments participants were asked to mentally 'see' photons going through one of the slits (the apparatus was locked within a metal box), or to block them etc. —something that would bring their conscious observation into one of the slits. Though more modest (for example, odds of 18 to 1 and 12 to 1) the results were indeed supportive of the hypothesis.

A further experiment attempted to test the strange time relationships occurring within the delayed-choice experiments; that is, whether observation could *retrospectively* collapse wave forms. You might recall in the description of the delayed-choice experiments, that the quanta had to communicate back in time with themselves once they reached a splitter which 'erased' their 'which path' information (information about the route they took to get to the splitter). They needed to do this to determine how they would go through the circuit—as a wave form, through all the possible (both) courses, or through a single course as a particle.

(The discussions and explanation of the delayed-choice quantum eraser experiment— in particular, the role of the coincidence counter—in Appendix 1 will help you understand the significance of the following experiment. If you were thinking about giving it a try, this might be a good time.)

In this remarkable experiment, Radin and his fellow researchers placed the double-slit apparatus by itself in a shielded chamber, with no one present in the laboratory, and recorded fifty sessions in April 2009. The data, however, remained unobserved.

In June of the same year, people were asked to take part in the experiment which, similar to previous experiments, provided computer audio signalling which fed them information on their success at collapsing wave forms into particles in real time.

A computer voice asked the participants, one at a time, at alternate intervals, to focus their attention on the double-slit apparatus, or to withdraw their attention and relax. The data that was presented to the participants was the data which was generated and recorded in April.

When asked to concentrate, the participants directed their attention towards the two tiny slits located inside the double-slit optical system. It was explained that this task was purely in the 'mind's eye', that is, an act of imagination. The test examined retro-causation, as it was testing whether the participants could change the results of a test which had already taken place and was recorded *some three months earlier*.

'...the mind is not in ordinary space-time in the first place.' Dr Dean Radin.

Twenty-two participants were recruited at a conference in Tucson, Arizona. The tests were conducted in an office at the conference hotel. After the test sessions were completed, twenty-two of the remaining unobserved data files were used as a control sample. Of the twenty-two participants, ten said they practised meditation regularly. This sub-group supported the hypothesis, recording significant results with odds against chance of 175 to 1. The non-meditators got chance results.[115]

Radin repeated the tests. In a series of six 'wave-function collapse' tests with the double-slit system, the combined outcome produced odds against chance of 184,000 to 1. Overall, the meditators performed better than non-meditators, with odds against chance of 300,000 to 1.[116] The control sessions provided no artefacts that might have been responsible for these effects. The tests were published in the journal *Physics Essays* in 2012.[117]

Radin wasn't the first to show how the mind can reach back in time and influence events. Theoretical physicist Helmut Schmidt, for example, has been collecting data on this effect since the 1970s. In 1988 he and medical researcher Marilyn Schlitz[xl] published multiple large-scale highly controlled experiments showing such effects. In one, a computerised random-generation process produced batches of 100 tones (sounds) of varying duration, some pleasing to the ear and some as bursts of noise. The random process was expected to produce roughly half of each.

Recordings were sent to volunteers tasked with listening to them and using their minds to try to increase the duration of the pleasant sounds while decreasing the duration of the noise. Once the participants completed their tasks they reported back to Schmidt and Schlitz, who then examined the original tapes. They discovered that the recordings the subjects had listened to contained significantly longer stretches of pleasing sounds than of noise.

In another test, subjects were asked to perform a similar task but to create more high notes than low ones. Again, the retroactive psychokinetic effect was found. These test produced outcomes that had a one in 8,000 probability of being the result of chance.[118]

Dunne and Jahn at the PEAR labs also undertook similar tests, some 87,000 'time displacement' experiments using random-generator computers, which also showed psychokinetic effects back in time. They commented that the effects are often greater in these experiments than regular psychokinesis testing done in real time.[119]

Schmidt and other scientists consider that retroactive psychokinetic effects are consistent with the observed outcome of a quantum process; that is, a process does not achieve a definite value or measurement until it is observed by a conscious being (even if a considerable period of time elapses before the observation takes place). It is for this reason that these experiments only seem to work when the initial recordings are not observed by people in the first instance, but observed after the psychokinetic efforts have been made.

---

xl Marilyn Schlitz is the former Director of Research at the Institute of Noetic Sciences (set up by Apollo 14 astronaut and sixth person to walk on the moon, Edgar Mitchell), and senior scientist at the Geraldine Brush Cancer Research Institute at the California Pacific Medical Centre.

Based on these and similar experiments, Radin comments that we should be 'used to the idea that the mind has access to information from "outside" the usual boundaries of space and time. This implies either that the mind can reach through space and time, or—and I'm guessing that this is more likely—the mind is not in ordinary space-time in the first place'.[120]

Presentiment

After his work with SRI International, where he conducted classified psi research for the US Government, Radin conducted *presentiment* experiments at the University of Nevada in the late '90s and at a prominent Silicon Valley think tank set up by Paul Allen (co-founder of Microsoft), Interval Research Corporation, between 1998 and 2000.

He used more advanced technology (including technology to monitor skin conductivity) and was able to show solid evidence that the body actually scans the future. In measuring the reaction of the body below the conscious level, such as through skin conductivity, he tracks how the body reacts physiologically, at a subconscious level, to different stimuli. In these experiments, he used pictures—some sedate, such as a picture of a lamp; others highly emotional, such as pictures of an act of extreme violence.

Using the technology, Radin tracked the reaction of the people to the pictures, showing that the body reacts differently to the more emotional pictures.

So far no big deal. However, intriguingly, the results also showed that the participants' bodies reacted to emotional pictures that were about to be presented, *before* the random-generator computer decided which picture to present, a confronting/emotional picture or a sedate one.

Over four experiments, Radin achieved results with odds against chance of 125,000 to one.[121] These results have been replicated by other researchers in different countries since.

At one point in the research, Nobel Laureate Kary Mullins took part in a demonstration. On a national radio interview a few weeks later, he explained to the audience that he 'could see about three seconds into the future ... There is something funny about time that we don't understand, because you shouldn't be able to do that'.[122]

Moreover, University of Amsterdam psychologist Dirk Bierman, who had successfully conducted several presentiment experiments, reasoned that if the effect was true, it could be recorded in experiments that were conducted for other reasons but were similar to the presentiment tests.

Bierman found three tests from which he could extract the data he needed. The first was a study of the speed with which fear arises in animal-phobic versus non-phobic people. The second was on decision-making, using a task known as the 'Iowa gambling task'. The third study investigated the effect of emotional priming on the evaluation of Japanese characters. For all three studies Bierman asked an assistant who was blind to the task to analyse the skin conductivity results of the research.

He found that in all three studies skin conductance behaviour had occurred as predicted *before* the stimulus. When combined, the studies showed a result that was significantly in agreement with the presentiment concept.[123]

Neuroscientist Julia Mossbridge conducted a similar meta-analysis with psychologist Patrizio Tressoldi and statistician Jessica Utts entitled *Predictive anticipatory activity preceding seemingly unpredictable stimuli: A meta-analysis*, published in *Frontiers in Psychology*. As Mossbridge recalls, 'we conducted a meta-analysis testing the unusual idea that human physiology anticipates what seems to be unpredictable events. This meta-analysis, which used statistically conservative methods to examine more than 40 studies published over the past three decades, found a small but highly significant overall effect in support of the hypothesis.'

'After examining several possible explanations for the effect, my co-authors and I concluded that we do not understand the mechanism underlying it. However, we are sure to state in the paper and the abstract that the mechanism is of course not supernatural (because, of course, no observations of the natural world can be beyond it). If the results are not due to fraud or other difficult-to-test explanations, they support the idea that retro-causality or the "backward" temporal flow of information can occur in physiological systems, at least at the subconscious level. This paper was cited by *The Wall Street Journal* Ideas Market, *Science Daily*, and ABC *20/20* websites, among others.'[124]

This research provides clear evidence that the body can retrieve information through a subspace-type medium to provide information about the future, in violation of space-time barriers. It provides tell-tale evidence that there is a part of us perceiving this information before we are consciously aware of it.

The significance of Radin's experiments

Radin's experiments with presentiment have remarkable implications. The computers randomly selected the images after the equipment had detected physiological changes in the test participants. In these tests the bodies of the participants are not predicting a likely future based on some kind of hyper-complex analysis of the world that their subconscious minds are performing in the background—their bodies appear to be perceiving a future that has absolute existence. This adds further weight to the work of Einstein and others, suggesting the existence of the future and past *with* the present.

It shows, like the observer experiments in quantum mechanics, that the future can inform the present, and change the present. In turn, this may even change the future.

It shows that the arrow of causality we experience in our everyday world is only one part of the story, supporting again what physicists are telling us, that there is no reason to see causality in only one direction. The ability of the future to constrain the present has also been shown in recent testing published in *Nature Physics* in 2012.[125]

These tests provide not only support for a subspace-type dimension that our bodies are in contact with, but further evidence that the present is actually the result of the future as well as the past.

Other results

The above only represents a slice of psi research. The body of research is too great to go into in depth here. Other examples include the following:

Card predictions

Radin's meta-analysis examined 309 studies, between 1935 and 1987, of the ability of people to predict which card was going to be chosen by a computer or a random process. The results could only be accounted for by chance as one in 10^{24} (ten million billion billion to one).[126] As with all the meta-analyses carried out by Radin they accounted for the possibility of a *file-drawer effect*; that is, the possibility that only the successful tests were recorded.

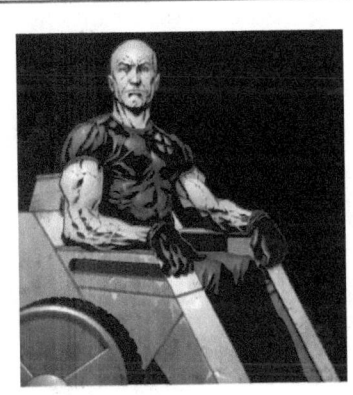

The character Charles Xavier or 'Professor X' was created by Stan Lee as the founder of the X-Men in the comic series. He has super telepathic abilities. His character is played by Patrick Stewart in the X-men movies. Effect added.

Training research

Experiments have confirmed that people can train themselves to become better at predictions. Charles Tart conducted experiments on a 'Ten-Choice Trainer', where he trained people to improve their telepathic ability. The test produced results that could only be explained by chance as having a probability of 2×10^{24} (two in more than a million billion billion).[127]

Military research

In a US Air Force report published as *Teleportation Physics Study*, its author writes about many scientific studies that suggest the teleportation phenomenon is possible at a macroscopic as well as at a quantum level. It refers to the work of quantum physicist Anton Zeilinger, who wrote a proof of quantum teleportation.

The report cites Chinese studies which were controlled, blind and double-blind and carried out in the Aerospace Medicine Engineering Institute in Beijing. The experiments were carried out with gifted children and young adults and proved that teleportation was possible across dozens of metres for small radios, photosensitive paper, insects and mechanical watches.

The tests were later repeated and the results of the teleportation recorded on video with extremely high-speed photography. Objects such as nuts, matches and live insects were moved from sealed envelopes and bottles. Videos showed the objects vanishing and appearing somewhere else, at times looking as though they merged with the walls of the containers. They showed that the objects were sometimes imperceptible, as if suspended in another location.[128]

Animals

According to a 2,400-year-old document, ancient Greeks observed animals such as snakes and rats abandoning the city of Helike before a devastating earthquake destroyed it. Throughout history and into the twenty-first century, reports of strange animal behaviour in advance of a natural disaster continue to inspire research and speculation.[129]

Modern-day studies of animals have found solid evidence of psi abilities, from pigeons finding their way home in impossible situations to scientific studies of dogs being able to predict when their owners were on their way home.[130]

Recently a research team led by UC Berkeley ecologist Henry Streby, studying the movements of golden-winged warblers, found that the birds were able to predict the approach of natural disasters—in this study, powerful tornados—and travelled long distances to avoid them. The study, published in the journal *Current Biology*,[131] showed that the birds fled while the storm was about 900 kilometres away, before atmospheric pressure and wind speed had changed. While meteorologists were only able to say that the storm would arrive at some time, the birds were already 'packing their bags and evacuating the area', Streby explained.[132] It was difficult for the researchers to explain the phenomenon, which they discovered accidentally. The best they could guess was that perhaps the birds could hear *infrasound*—acoustic waves at frequencies below 20 Hertz (cycles per second), which may be caused by natural disasters. These frequencies can't be heard by humans.

Radin describes research conducted in a master's thesis in electrical engineering by Chester Wildey. His research was able to show that earthworms could predict when, in presentiment fashion, mechanical vibrations were going to occur in soil.

Fernando Alvarez of the Estación Biológica de Doñana in Seville, Spain, placed forty-seven Bengalese finches into a cage. After allowing them to become acclimatised to the cage, a random-number generator selected a random time to display a fifteen-second video clip on a screen next to the cages. The clip contained a horseshoe whip snake crawling towards the birds, and made the birds visibly distressed when played..

The birds were continuously filmed during the experiments and the results were compared with two control groups. The results showed that the birds reacted to the snake video clip up to nine seconds before it was shown, with odds against chance ranging from 5,393 to 1 compared with one control, and 280,000 to 1 compared with the other.[133]

Again, the above is only a selection of the research that has been done confirming psi phenomena.[134]

Analysing psi evidence

Psi results have become so convincing that they are not questioned so much anymore, not even by the more outspoken sceptics. Recently, some of the most prominent sceptics admitted, for example, that ESP had passed the test of science.[135] Most now admit *something is happening;* but *what*, no one is clear about.

Radin considers that Hollywood movies have made psi abilities less legitimate in the eyes of the scientific community. He provides examples in psychological texts and sceptical websites of claims that psi phenomena are either fraudulent or that most studies in respected journals have showed negative results. As someone who has made a profession out of examining research in this area, Radin comments:

'The truth is that in the entire history of laboratory investigations of these phenomena, spanning more than a century, there is one case of proven fraud and two or three suspected cases. Compared to the number of cases of scientific misconduct in conventional disciplines, ranging from outright data fabrication to duplicate publication and plagiarism, this domain is positively saintly ... The plain fact is that there is no body of "overwhelmingly negative" results.'[136]

Radin writes of one study, in 2008, when newspapers around the world hailed what was called the first conclusive test for telepathy, conducted by two Harvard University researchers. It used high-tech brain scanners (fMRIs) to detect neural evidence of ESP 'once and for all'. It found no evidence.

Radin said it felt conclusive until one read the article, which showed one of the sixteen tests achieved a stupendously significantly positive outcome. More so, Radin found that it wasn't the first test to use MRI machines, or the second, but the seventh. The six others, all conducted since 2000, showed highly significant evidence in favour of psychic abilities. Somehow the researchers didn't feel it important to mention these other tests, and the newspapers didn't bother finding out. Why? As Radin suggests, 'the study confirmed their prior beliefs, so no one felt it was necessary to check the facts'.[137]

Radin provides his own site, a *show me* page with downloadable articles on psi and related topics, all published in peer-reviewed journals, most published after the year 2000. Most report experimental studies or meta-analyses of classes of experiments, including studies related to healing at a distance, physiological correlations at a distance, ESP, precognition, presentiment and mind-matter interaction (*http://noetic.org/research/psi-research/*).[138]

What is clear from this and other research is that the balance of evidence does favour the existence of a dimension which our mind is able to access, but which is difficult to discern through the four-dimensional reality (three special dimensions plus time) that we have become so accustomed to. It connects people and objects across distance and time in contradiction to our everyday experience.

Max Planck, the discoverer of the quantum, wrote that 'there are realities existing apart from our sense perceptions'.[139] We appear to be like islands in the sea, with an above-sea-level view, unable to see the land underneath that connects us.

Research into our unique *sixth sense* abilities is not new and has been conducted by some of the most respected scientists since our move out of nineteenth century naturalism, including Nobel laureates.[140] Albert Einstein even wrote the preface for a book by Pulitzer prize-winning author and social activist Upton Sinclair, called *Mental Radio* (1930), showcasing the results of experiments supporting clairvoyance.

While the above mass of modern scientific evidence provides very convincing evidence for our abilities in this area, the scientific establishment of today still has difficulty giving it the attention it appears to deserve.

Others beside Radin argue that this is the result of a materialist philosophy that has a firm grip on the purse strings of scientific research,[141] sustained by a capitalistic motivation to have science focused predominantly on producing sellable merchandise, like a new iPhone, a new medicine or a new fighter jet.

There are always going to be good ideas, and it is understandable that research dollars in our modern times are more likely to be directed towards areas where a financial return can more easily be expected. Equally, many are resistant to the idea that you can buy the answers to the big questions.

One can understand how difficult it can be to conduct research purely for knowledge's sake, in areas difficult to patent, package and sell at a decent profit, and how this is a slow and persistent force keeping areas like psi out of the scientific mainstream. It is unfortunate that some spectators, less-versed but eager to please, make the mistake of interpreting psi's minority status as a sign of its inferior science. Perhaps researchers like Radin should take comfort in the words of Jonathan Swift, 'when a true genius appears in this world, you may know him by this sign, that the dunces are all in confederacy against him'.

NEAR-DEATH EXPERIENCES (NDES)

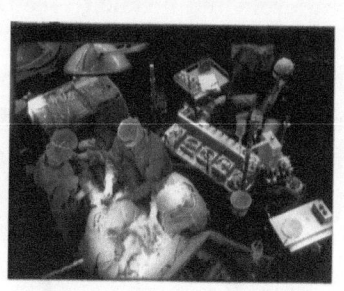

Brain-dead patients have been known to rise above their bodies and observe what is happening around them. This is a common feature of a near-death experience.

The NDE is a profound experience. It is often an out-of-body one that takes place when a person's brain has stopped functioning or when they are confronted with a life-threatening situation. Reality is experienced from another point of view, outside everyday dimensions, where gravity, time and space have different properties. It is often accompanied by spiritual experiences.

Pim van Lommel is one of a number of medical and health practitioners who have been drawn to NDE research.[142] A world-renowned cardiologist, Lommel was struck by the commonality of NDEs reported by his patients. The experiences included lucid consciousness while brain-dead. Some were able to recall real events that occurred during their unconscious state from a perspective outside their bodies. Some experienced tunnels, bright lights, reviews of their life, and a timeless state where they would meet people already deceased. Deaf people talked about what they *heard* and blind people, who had never experienced sight, gave vivid accounts of what they *saw*.

As research later confirmed, a significant number were able to have their recollections of events corroborated by independent witnesses. In one survey of 93 out-of-body experiences described by Lommel, 43 per cent of reports had been corroborated by independent witnesses who were available to the researchers at the time of the surveys. eighty-eight per cent of these were completely accurate. Only 1 per cent were completely erroneous.[143]

Lommel started his own serious research into these experiences and has produced some ground-breaking results. Since his initial study, published in the prestigious medical journal *The Lancet*, Lommel has resigned as a practising cardiologist to focus on this research.

Lommel describes how this has dramatically changed his views about existence and science. 'That death is the end, used to be my own belief,' he writes. 'But after many years of critical research into the stories of the NDE-ers, and after a careful exploration of current knowledge about brain function, consciousness and some basic principles of quantum physics, my views have undergone a complete transformation. As a doctor and researcher, I found the most significant finding to be the conclusion of one NDE-er: "Dead turned out to be not dead." I now see the continuity of our consciousness after the death of our physical body as a very real possibility.'[144]

In his book *Consciousness Beyond Life*, Lommel discusses his and other research internationally. He comprehensively surveys the common everyday explanations put forward for the NDEs, such as drug-induced hallucinations, or an oxygen deficiency causing visions and dreams. He provides medical and scientific reasons why these explanations are unsatisfactory.

He supported his findings with a large-scale longitudinal study (repeated observations over a long period of time) with psychologists in the Netherlands in 1988 (referred to as the 'Dutch Study', published in *The Lancet* in 2001). Over a four-year period, between 1988 and 1992, 344 consecutive patients who had undergone 509 successful resuscitations were included in the study.

All patients had been clinically dead as a result of a lack of oxygen supply to the brain (anoxia), during which the brain stopped functioning (Lommel provides a comprehensive medical explanation of why the brain stops functioning[xli & 145]).

The study correlated the experiences of patients with the common elements of NDEs, developed in 1975 by psychiatrist Raymond Moody. Moody describes twelve elements of NDE but emphasises that most people experience only a few. They include the ineffability of the experience—what happens in the NDE is often unfamiliar and indescribable, lying outside the sphere of normal experience. They also include a painless feeling of peace and quiet; the awareness of being dead; an out-of-body experience; the experience of a dark space; a tunnel; sometimes a fearful event; meeting deceased persons; a brilliant light or being of light; a panoramic life review; the perception of a fast-forward review of part of life yet to be lived; the perception of a border or threshold; and the conscious experience of a return to their bodies.[146]

The research showed that these elements were experienced consistently, and that a person's religion, level of education, fear of death, medication or duration of unconsciousness were irrelevant to the experience of an NDE.[147]

The study found that the possible causes of NDEs—physiological or medical causes (such as anoxia), psychological (such as fear of death) or pharmacological (causes related to medication)—could not be corroborated.[148]

The follow-up interviews of the participants also confirmed that those who had experienced NDEs had significantly different and sustained changes to their approach to life from those who did not expe-

---

xli Brain studies have shown that under normal circumstances a functioning, collaborative network of brain centres is a prerequisite for the experience of waking consciousness. This is absent during cardiac arrest. Additionally, there appears to be scientific proof that the cerebral cortex and brain stem are devoid of measurable activity during a cardiac arrest and that the clinical picture also reflects a loss of all brain function.

rience an NDE during cardiac arrests, including a reduced sense of fear of death and a greater level of spirituality.

Similar studies in the US and Britain among cardiac-arrest patients, with the same design as the Dutch study, produced nearly identical results.[149] These were conducted in the US by Bruce Greyson, a professor of psychiatry and neurobehavioral sciences, and in the UK by intensive-care physician Sam Parnia, neuropsychiatrist Peter Fenwick and Penny Sartori, a senior intensive-care nurse. The studies comprised 562 patients. They showed the same percentage of NDEs during cardiac arrests, no physiological or psychological explanations for the NDEs, the occurrence of NDEs during the cardiac arrest and the cardiac arrest resulting in the loss of all brain functions.

In his conclusion, Greyson writes, 'No one physiological or psychological model by itself explains all the common features of near-death experiences ... The paradoxical occurrence of heightened, lucid awareness and logical thought processes during a period of impaired cerebral perfusion (blood flow to the brain) raises particularly perplexing questions for our current understanding of consciousness and its relation to the brain function ... A clear sensorium and complex perceptual processes during a period of apparent clinical death challenges the concept that consciousness is localised exclusively in the brain.'[150]

Relationship between the brain and cognitive functions

The research into NDEs exposes a difficulty science has had for some time in understanding the role of the brain in supporting thought and memory processes—the hypothesis that consciousness and memory are produced exclusively in the brain remains unproven to this day.[151]

We have seen above how tests done by Sheldrake and others provided evidence that the mind was not confined within the skull. But also, to date, science has not explained how neural networks produce the subjective essence of thoughts and feelings; so far no neurophysiological study has identified any exact correspondence between specific neural activities and the specific content of memories, experiences, feelings or thoughts.[152]

While modern technology such as EEGs, fMRI and PET scans have been able to show correlations between activities in the brain and experiences, these activities can occur throughout the brain, and sometimes in remote areas of the brain simultaneously. Scientists understand that correlations say little or nothing about cause and effect; for example, someone can activate a radio and tune in to a station playing an Elton John song, but this doesn't mean Elton John is inside the radio. Nor does it mean that what we do with the radio dials determines the words of the Elton John song; movement in certain parts of the brain does not imply that those areas create the thoughts, any more than electric signals in a telephone line mean that the telephone line creates the signals.

There are claims that electrical stimulation of the brain can produce certain mystical experiences,[xlii] including spiritual ones. However, most of us can appreciate that just because one can electrically stimulate the brain to make a person see the colour blue,[153] doesn't mean that the photons that normally provide us with the experience of the colour blue don't really exist.

Indeed, science has shown that electrical stimulation as well as medicine can be useful therapeutic tools. But what is even more interesting is that placebo treatments have been able to produce the same neurological improvements in the brain.[154]

[xlii] Michael Persinger of Laurentian University in Ontario claims to have created religious experiences in the brain through electrical charges from his 'God helmet', which generates weak electromagnetic fields and focuses them on particular regions of the brain's surface. Although a 2005 attempt by Swedish scientists to replicate Persinger's God helmet findings failed, researchers are not yet discounting the temporal lobe's role in some types of religious experience.

MIND TO MIND, INTER-DIMENSIONAL COMMUNICATION

'The whole of history of science shows us that whenever the educated and scientific men of any age have denied the facts of other investigators on a priori ground of absurdity or impossibility, the deniers have always been wrong.'[55]

Alfred Wallace, 1893.

If we are capable of perceiving this other dimension, then it would appear at least theoretically plausible that we might be able to access these dimensions in other situations that don't require our brains to be dead or experience a dramatic near-death event. If our conscious self can indeed survive death, as the NDE research suggests, then what prevents us now from perceiving the consciousness of others, through subspace? Is there evidence that we can do this?

Witches, snake-oil salesmen and mediums

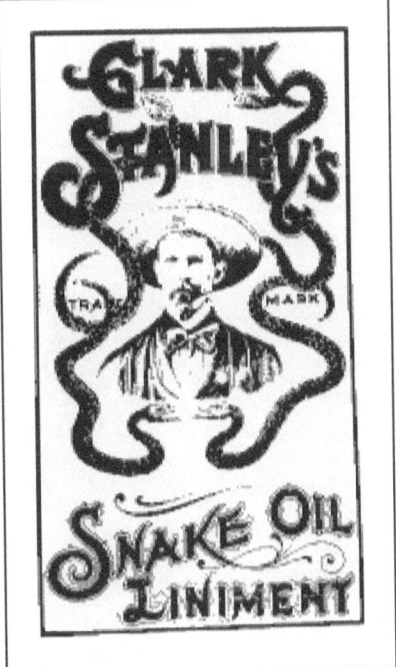

'Snake oil' is an expression that once referred to fraudulent health products or unproved medicine, but now is used to describe any product with questionable benefit. A 'snake-oil salesman' is someone who knowingly sells fraudulent goods or who is a fraud, quack or charlatan.

It is believed by some that the term came from the Western regions of the United States and is derived from a topical preparation made from the Chinese water snake, used by Chinese labourers to treat joint pain.[156] It is thought that North American travelling salesmen used the name to sell all sorts of medicines of dubious value as a cure for everything. To get people to buy the product, they sometimes placed stooges in the audience to claim that the medicine cured an illness. The salesman would typically leave town before the medicine's real worth was discovered.

There were also times when knowledge of medicine or natural therapies was interpreted as witchcraft and its practitioners were tortured or burnt at the stake. Feminist historians, for example, claim that women persecuted as witches in Europe and America during the sixteenth and seventeenth centuries had been the traditional healers and midwives of their communities. Their destruction had not merely been a blow against the power of women in society but against natural medicine and therapies. The witch trials were therefore thought of as a victory for both patriarchy and a flawed, male-dominated modern science.[157] During the hysteria of the witch-hunts some think as many as 100,000 people, mostly women and children, were executed.[158]

Despite the fraudulent salesmen and hysterical superstitions, our exploration of medicine did continue. It has become more regulated and now medicine and natural therapies are widely accepted around the world as a legitimate form of treatment for illnesses. Those who practise medicine are also considered honourable people and I can't even recall the last time someone was burnt at the stake around here for their healing skills or concoctions. Though, tragically, children and women are still murdered for being thought to be witches in some parts of world, such as Sub-Saharan Africa.

Mediumship is also vulnerable to fraud and superstition. In this context, a *medium* is a person who acts as a spiritual intermediary between the dead and the living.

Spiritualists often set March 31, 1848, as the beginning of their movement. On that date, Kate and Margaret Fox, of Hydesville, New York, reported that they had made contact with a spirit. It is said that for payment at least one of them recanted their stories later in life.

History and folk law abound with fraudulent claims of mediumship and hunting the phoneys is good entertainment, even today. For some it provides all the thrills of an old-fashioned witch-hunt.

Authorities at times have wanted to crack down on the practice. The *Fraudulent Mediums Act 1951* was a law in England and Wales which prohibited a person from claiming to be a psychic, medium or other spiritualist while attempting to deceive and to make money from the deception (other than solely for the purpose of entertainment).

This act repealed the *Witchcraft Act 1735*, which was partly brought into being to stop the local execution of people accused of witchery, not only for their healing practices but due to beliefs that they could communicate with the spirits of the dead. At the time this was considered by some as an evil in and of itself.

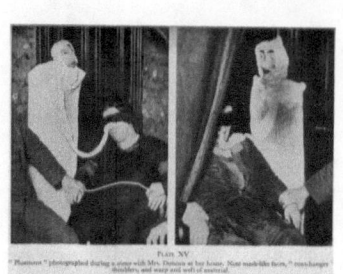

Photographs said to reveal the fraudulent mediumship of Helen Duncan, a medium in the UK. Malcolm Gaskill revealed in his book HELLISH NELL: LAST OF BRITAIN'S WITCHES *(Fourth Estate, 2001)* that the photographs were taken by the photographer Harvey Metcalfe in 1928 during a séance at Duncan's house.

The *Fraudulent Mediums Act 1951* was repealed in 2008 by new *Consumer Protection Regulations* following an EU directive targeting unfair sales and marketing practices.

While tabloid newspapers are quick to portray a stereotype of the fraudulent medium preying on the vulnerability of those struggling with the deaths of their loved ones etc., rarely do they take the time to explore the actual science that is building around mediumship. Mainstream articles generally don't go any deeper than portraying mediums as charlatans or, at best, pathologising the ability, suggesting that mediums are mentally ill Good Samaritans.

One of the difficulties mediumship has is that it appears to be a little more like going fishing than manufacturing. Sceptics get angry with mediums because they often can't produce answers to specific questions on demand. But to be fair, it is a bit like asking someone who is rod fishing off the beach to pull out an Atlantic salmon on demand. The phenomenon seems to depend on what is happening on the 'other side' as much as what is happening here—much like fishing depends a lot on what is happening under the water, largely out of the control of those fishing on the beach. You can't just order up what you want when you want it, and this is difficult for some to deal with. It also requires much more sophisticated research skills to study.

While there have been (and will continue to be) fraudulent mediums, there are examples of people who may well have such profound abilities. History has shown them to be used by some of our most astute leaders, including Queen Victoria, Abraham Lincoln, Winston Churchill and Charles De Gaulle.[159]

An intrepid breed of scientists has applied the scientific method to this phenomenon. As mentioned earlier, Alfred Wallace, one of the most renowned scientists of the nineteenth century subjected spiritualism to scientific study and supported his findings despite the damage it did to his professional career.

Others have continued in Wallace's footsteps and this has led to a small but growing acceptance within science that there is something much more interesting going on behind the headlines.

People with profound abilities are not common. You have to take advantage of the Michelangelos and Mozarts while they are here. Bertha Harris seems to have been one such person, born around the turn of the twentieth century, in Chester, England. She was known as a profoundly gifted medium, being called on by some of the most astute of her time to provide her services. She was a regular at number 10 Downing Street, reading for Winston Churchill during World War II. General Charles De Gaulle and King George of Greece also reportedly used her services during the war.[160] She toured England with Sir Arthur Conan Doyle and was known to have converted all the renowned sceptics of the day who put her to the test.[161]

Leonora Piper was studied by some of the leading scientists of her time.

In the US, Leonora Piper was 'discovered' by William James, a pioneering Harvard University psychologist, three years after the founding of the Society for Psychical Research (SPR) in London in 1882.

Piper was a young Boston, Massachusetts, housewife who delivered messages that seemed to be coming from spirits of the dead.

Soon after the discovery of Piper, the American branch of the SPR (ASPR) was formed under the guidance of Professor James, and its primary task became the study of Piper's mediumship, although it undertook the investigation of other mediums and paranormal phenomena as well. A number of other reputable scientists and scholars studied Piper for a quarter of a century.

RICHARD HODGSON, LL. D.,
BOSTON, MASS.

An Australian, Richard Hodgson, known as one of the most stalwart sceptics of the day, at one stage took over the testing of Piper. An esteemed academic at the time, James Hyslop, a professor of logic and ethics at Columbia University, also took control of the experiments for a period. They conducted sittings with Piper on average three times a week for about eighteen years.

Sir Oliver Lodge was regarded as one of the brightest English scientists of his time. He was made professor of physics at the age of 30, and was knighted and made a fellow of the Royal Society in 1902. He was known for his investigations of lightning, the voltaic cell and electromagnetic waves. There is a plaque at Oxford University on the spot where he sent the first radio message. He is known as the first person to transmit a wireless signal. He also travelled to the US to take part in Piper's testing.

After years of testing by believers and non-believers, all of the above-mentioned scientists became convinced of the genuine abilities of Piper. The once most sceptical of them all, Hodgson, confirmed, '... at the present time I cannot profess to have any doubt but that the chief "communicators" to whom I have referred in the foregoing pages, are veritably the personalities that they claim to be, that they have survived the change we call death, and that they have directly communicated with us whom we call living through Mrs Piper's entranced organism.'[162]

Sir Oliver Lodge stated, 'I have assured myself that much of the information supplied by Mrs Piper during trance has not been acquired by ordinary everyday methods and precludes the use of the normal sense channels.'[163]

Piper's abilities, however, were an affront to the prevailing scientific view at the time, very much entrenched in the classical, naturalistic paradigm. They were also denounced by some sections of religious orthodoxy (who believed that her communications were inherently evil).

Sceptics considered her a victim of hallucination and delusion and just another charlatan; though a really clever one, considering how many intelligent men and women she duped in hundreds of observations over three decades.

There are reports of occasions when Piper gave incorrect information,[164] suggesting to some that she was a fraud. Some tricked her, asking her to contact fictitious persons, and criticised her when she claimed that she did so.[xliii] A perusal of the Internet will find such examples.

There is still some controversy about her, even though it is claimed sceptics were not able to discredit her. Desperate to account for her positive results, some even resorted to the curious accusation that she was telepathic and only pretending to be a genuine medium for the dead.[165]

In his recent book, *Resurrecting Leonora Piper: How Science Discovered the Afterlife*, Michael Tymn provides some clarity on Piper's abilities. While acknowledging the inaccuracies of some of her readings, he suggests that those who carefully study the research will likely see Leonora Piper as the 'white crow' that William James proclaimed

---

xliii One investigator reportedly invented a dead niece whom he named Bessie Beale, and requested Mrs Piper's control to contact her spirit. Messages from the non-existent 'spirit' were duly given—Simeon Edmunds. (1961). *Hypnotism and Psychic Phenomena*. Hal Leighton Printing Co. p. 122.

her to be—the one who proved that all crows are not black, the one who gave science some very intriguing evidence about the afterlife.[166]

Evidence of genuine mediumship is not confined to modern history. Today, widely regarded as history's most important scientific investigation of the evidence for life after death, are the Scole experiments.

The Scole experiments

In *The Scole Experiment, Scientific Evidence for Life after Death*, Grant and Jane Solomon recounted the five years of experiments undertaken to study the Scole group's experiments with mediumship.

The investigations began in 1993 through the work of psychic researchers at the house of Robin and Sandra Foy in the small English village of Scole in Norfolk. Two subjects, often the Foys, entered a trance in the basement of their home, usually in total darkness, and became the mediums for several observers/investigators. They would become channels through which a team of individuals in the spirit world—apparently engaged in a project to develop techniques to communicate through dimensions—would communicate with them. The investigators would set up various apparatuses in order to facilitate greater communications with the other side, mostly on the advice of the group on the spirit side.

The phenomena they were able to show included direct contact with spirits, the materialisation of objects onto a table *(apports)*, the appearance of spheres of light that moved through solid objects, the creation of drawings, written communications and pictures on photographic film, the playing of musical instruments, the levitation of tables and the hearing of voices, some from within walls.

Towards the end of 1996, prominent researchers were invited to these sessions, including academics associated with the Society for Psychical Research: Robert L. Morris, Donald West, Archie E. Roy, Bernard Carr, Alan Gauld and John Beloff.

Other investigators, from a wide variety of disciplines and backgrounds, also attended the experimental sessions. These included Ivor Grattan-Guinness, Rupert Sheldrake, Ernst Senkowski, Hans Schaer, Kurt Hoffman, Russell Targ, Marilyn Schlitz and Bernard Haisch. Many of the independent researchers had wide experience with paranormal investigations and were familiar with fraudulent techniques.

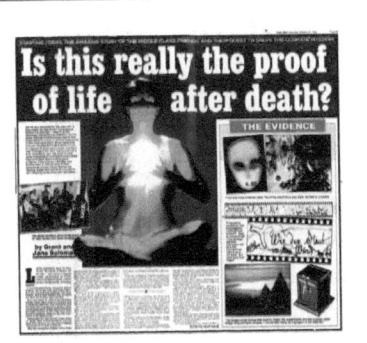

Article showcasing evidence from the Scole Experiments, TIME OUT MAGAZINE, *London 1999.*

The investigations also involved people from organisations such as NASA, the Institute of Noetic Sciences, the Scientific and Medical Network and Stanford University. The group also demonstrated their work outside of Scole, in locations around the world—including Germany, Ireland, the Netherlands, Spain, Switzerland and the US—where other investigators assessed and reviewed the work, including Ulf Israelsson, Hans-Peter Studer, Theo Locker, Andreas Liptay-Wagner, and Paul Kurthy. Some experiments were attended by celebrities and magicians.

Despite the host of phenomena produced all around the world, no one provided any evidence to suggest fraud in any form in the five years of the experiments.

For some investigators, the Scole group's findings appeared, in many respects, as the most important development encountered in the enduring effort to demonstrate the survival of human consciousness after the death of the material body.

One testimony was from Hans Schaer, a Swiss lawyer and businessman who owned a holiday home on the Mediterranean island of Ibiza, who had sat with the Scole group on a number of occasions. He invited them to demonstrate the phenomenon at his holiday home in 1995. This is his recorded testimony:

'I am a down-to-earth person, a die-hard realist and businessman with, due to my legal studies, a very critical and analytical mind. I am not psychic. All my life it has been my intention to find out—if possible—whether there is life after physical death.

'My research has involved visiting the Scole Hole [the Foys' house] on various occasions and I have participated in certain film and videotape experiments. I have personally conducted some experiments which have taken place under test conditions. I have been witness to a number of highly interesting phenomenon.

'I invited the Scole group to my old country house on the island of Ibiza. If they had ever faked anything within their own cellar, they had no chance whatsoever to do this in my home ...Just before one experimental sitting, I came up with the idea of asking the spirit team if they could provide "evidence" by playing a musical instrument. The Scole group had neither the opportunity nor the time to prepare anything before the sitting began.

'The result of this request was fantastic. The trumpet which I had placed on the table started playing, even though the mouthpiece was removed, and later on someone else started playing a drum solo on the wooden table, despite the fact that there were no drumsticks or other suitable objects available.'[167]

In the later years of the experiments, the experimenters found that people who had died were not the only entities that were communicating. They started receiving messages from entities claiming to have never been to Earth, who gave the impression that they were communicating from a far-off dimension. Some photographic evidence was also produced.[168]

A weakness of the research was that the controls and experimental scenarios were, for the most part, prescribed by the spirit group on the other side. Requests by experimenters on this side to change procedures to increase the scientific rigor of experiments was at times rejected by them, for reasons they did not make absolutely clear.

In the words of one researcher, Arthur J. Ellison, 'we were often told by the communicators that we would not be able to understand the explanation of what was going on. I think there were times when we all wished that they would just explain fully and let us decide whether we understood or not! But it was not to be.

'Also, we often explained that the scientific community would assume that, as the phenomena usually occurred in the dark, we were being deceived. The use by us of an infrared viewer showing, from their own body heat, that everyone was in their chair during the periods when the phenomena were taking place was highly desirable. However, this was not, to our regret, allowed either.

'We made particular efforts in our SPR report to explain how impossible faking many of the phenomena appeared to be. But the sceptic will always say that magicians can do all sorts of "impossible" things. In this subject, sadly, the sceptic is not usually required to demonstrate how what they claim could have actually been executed without detection.'[169]

It was curious that Professor Ellison's forward to the book finished on his personal opinion that 'the group on "this side" were honest and genuine. After two years we knew them extremely well. I consider that the results of the sessions were of great interest to science.'

Reading through the book one gets the impression that, indeed, something astonishing had happened. To conclude that it was all a hoax would require a conspiracy theorist of greater imagination and conviction than me. Besides, if these people were frauds, they were obviously incredibly skilled ones, and would likely make a whole lot more money as magicians than as spiritual researchers, an occupation with virtually no financial rewards. Though reading through the evidence suggests that a complete picture was being withheld—not by people on 'this side', but the spirit side. Naturally, this makes me uneasy, and makes me wonder whether the entities on the spirit side are always who they claim to be.

Other testing

Modern science doesn't stop at the Scole experiment; testing has well and truly moved into the laboratory, controlled by scientists 'on this side'. Among them is Gary Schwartz at the University of Arizona. Schwartz and his colleagues conducted a number of research studies with some of the best-known mediums in the United States, such as

John Edward, Laurie Campbell, Sally Morgan and Allison DuBois (on whom the television series *Medium* is based). After extensive testing, Schwartz wrote: 'These mediums have been tested under experimental conditions that rule out the use of fraud and cold-reading techniques commonly used by psychic entertainers and mental magicians.'[170]

Allison DuBois is an American author and medium. Her use of psychic abilities to assist US law-enforcement officials to solve crimes was the basis of the TV series Medium.

Personally, I have done my own field research of mediums here in Australia. I have had mixed experiences and can't say I have had 'my socks blown off'.

I have not yet encountered mediums who could contact the spirits of those who will live and die in the future. I am puzzled by this as we are told time behaves differently in the subspace realm? As so much of the advice coming from mediums relates to what the future holds, there doesn't appear to be a taboo on providing future-related information.

This is an interesting question, as the possibility of hearing from spirits who have died in our future brings with it the possibility of communicating with *oneself*, in spirit form. If anyone knows the answer to this question, I'd like to hear from you.

As serious research into mediumship continues, I'm sure questions like the above will be answered. Science is becoming more aware of the fundamental properties of this subspace realm and this presents important opportunities to explore connections between medium-research and the rest of science, to the advantage of all.

If any of the above is true, it appears that information is coming from or through a subspace-like realm. I suspect this world could be quite different in nature to what we imagine (or perhaps are led to imagine), or even what we are *able* to imagine; that is, 'queerer than we can suppose'. This may also be the case for the entities that reside in that world.

SUMMARY

Our neocortex has provided humans with unique pattern-recognition skills, enabling us to perceive higher-order contextual information. Evidence is increasingly suggesting that the brain is analysing information from the world to inform *something* that most often is defined as *consciousness*. The idea of higher-order realities that can be understood and acted upon through consciousness was all but thwarted by the nineteenth century naturalistic view of reality, perceiving it as a threat to a bottom-up deterministic, mechanical view of existence.

While the naturalist approach has given us technological advances, it ironically has paved the way to the discovery of quantum physics, which turned our understanding of reality upside down. The sanctity of the physical laws, the absolute nature of space and time (often termed *locality*), and the impotence of our experience of consciousness were all undermined by what is now widely recognised as the most tested area in science, quantum mechanics.

Contrary to the classical view, it has been shown that consciousness can play a role in assisting our understanding of life and its unfolding. It can account for the very knowledge we are able to generate, allowing us some degree of independence from the chain reaction of determinism. It can explain some anomalies which naturalistic approaches can't, such as examples of top-down causation, including the placebo effect, the 'gatekeeper' role of consciousness, the incredible results of psi research (such as telekinesis, clairvoyance and telepathy), and our abilities to bring ideas into the space-time reality (from a sculpture to a 747). Indeed, it can even play a role in creating a more elegant and complete theory of evolution.

Importantly, our consciousness and mental processes appear to have at least a foothold in the primal subspace dimension, a dimensional footing which allows us some freedom from the chain reaction of space-time. It allows us to observe, gather knowledge and ultimately interfere in the universe. As shown, this interference is the immaterial mind directly controlling the material body

The independence of consciousness from space-time is further demonstrated in other living experiences, including near-death experiences and the phenomena of mediumship. Mind-to-mind communication appears to be a reality we need to get over. Keeping our prejudices and superstitions under control will help us identify the science behind these phenomena and the role of the mental as well as physical in the unfolding of life and what appears to be a more enduring phenomenon, what many refer to as their *souls*.

This also raises another question: if the subspace reality has been so much a part of our existence until now, how have we been able to deal with it? What has our experience taught us? What do we already know?

Religion has been slowly separated from institutionalised science over the past two hundred years. The modern institutionalised manifestation of both makes them look quite different, but for those who know their history, these differences can look rather like facades. At a fundamental level, separating knowledge into religion and science is a little like unscrambling an egg. To understand either requires the understanding of the development of both.

Many feel that the institutionalisation of this separation has neglected important connections and skewed both sides. This has included the lessons of religion being leached slowly out of mainstream Western education and again many of us are seeing the effects of this. Perhaps it has resulted in us trying to solve our complex modern problems with one hand tied behind our backs.

In gaining knowledge about a cosmos as profoundly complex and mysterious as ours, many are realising the downside of being too particular or fussy about who we want our messages delivered by. In order that we benefit from both institutionalised science and institutionalised religion in building our knowledge for the future, we need to understand why religion should be invited back to take its place at our table of knowledge. Both sides have much to gain, but humanity has even more.

CHAPTER 4
Religion, the elephant's head

RELIGION, THE ELEPHANT'S HEAD

'The first gulp from the glass of natural sciences will turn you into an atheist, but at the bottom of the glass God is waiting for you.'[1]
   Winner of the Nobel Prize in physics, Werner Heisenberg.

Religion has been defined as 'a set of beliefs concerning the cause, nature, and purpose of the universe, especially when considered as the creation of a superhuman agency, usually involving devotion and ritual observance, and often containing a moral code governing the conduct of human affairs'.[2] What does a review of scientific research tell us about religion's insights into subspace?

It is impossible to understand religion and its human experience without understanding its subjective element. Known as the father of modern psychology, William James suggested that one of the key characteristics of religion is its ineffability—its inability to be adequately expressed in words, something which needs to be experienced to be understood. 'No one can make clear to another who has never had a certain feeling, in what the quality or worth of it consists.'[3]

For some, this understandably is a sign that perhaps there is nothing behind the explanation or experience. If it, and phenomena such as near-death experiences, can't be clearly articulated, then perhaps it is because there is nothing real behind it. While this seems plausible, the logic breaks down rather quickly when one tries to figure out how to explain the colour blue to a person who has never had the experience of sight, or how to explain hearing to a deaf person. Does our difficulty explaining sounds mean that sound waves don't exist?

I must declare, I am not the most religious person around. The closest I get to organised religion these days is at marriages and funerals. While I do have my own spiritual beliefs I find myself less religious around judgemental theists, and the reverse around smug atheists. As one who has been both in the past, I understand how much these polar opposites have in common.

Like many, I am also aware that science isn't alone in falling short of its own principles. The same can be said for religion. The failures of institutionalised religion to live up to its own moral codes go beyond the child abuse, Taliban and Islamic State stories that invade headlines today, to the extent that it would be trite to list them here. Religion has not been immune to the faults of society; religion has always been a part of society in some form. Nevertheless, there is no denying religion's fundamental role in human history as a force for civilisation and, for want of a better word, *goodness*—what many of us intuitively understand to have importance.

SO WHAT IS RELIGION DOING AT THIS SCIENCE FAIR?

'I maintain that cosmic religious feeling is the strongest and noblest driving force to scientific research.'

Albert Einstein. *Religion and Science, 1930.*

There are various theories about the 'evolution' of religion. Many of them focus on the development of the mental attributes that incline humans to religious beliefs, such as our ability to understand complex causality, our social attributes, our mystical experiences and the emergence of our so called *existential anxieties*—our awareness of our own mortality and self-suffering.[4]

Archaeological evidence of religious-like symbolism can be seen as far back as a million years. However, it is estimated that the arrival of the anatomically modern human in Africa about 150,000 years ago through to the 'sudden' arrival of culturally and technologically sophisticated humans in Europe from about 40,000 years ago is likely to be the period within which the mental foundations of modern religion were laid.[5]

It is pretty well acknowledged across the educated community that religion is a major influence in the creation of civilisations, from the justice system to health care, human rights and the arts.[6]

Yet many would be surprised to learn that science actually sprouted from religion. The word *science* denotes the structured process of generating knowledge through observation, theorising and testing, practised since humans came down from the trees, defined by the ancient Greeks and formalised by Muslim scholars as the *scientific method* during the Islamic Golden Age. Medieval religious writings acknowledged the essential relationship between the study of nature (science) and religious thought.[7] Indeed, Sir Isaac Newton is likely to have considered himself more of a theologian than a physicist, due to the extent of his theological writings (more later).

It is claimed that the Dalai Lama said that if he had to choose between science and religion, he'd chose science. He regularly hosts discussions between Buddhist monks and Western scientists. 14th Dalai is the current Dalai Lama, as well as the longest lived incumbent. Dalai Lamas are the head monks of the Gelug school, the newest of the schools of Tibetan Buddhism. He won the Nobel Peace Prize in 1989. Source: Photo by Luca Galuzzi.

The term *scientist* was actually created to be an umbrella term for people undertaking research in many different knowledge-gathering fields.[8] It was a term invented to be inclusive, rather than exclusive, in contradiction to how some corners of modern science use the term today. As Philip Clayton, author and Dean of Faculty at Claremont School of Theology, put it, it is 'possible to be scientific about one's religion and to be religious about one's science'.[9]

The Dalai Lama regularly hosts discussions between Western scientists and Buddhist monks. Scientists compete for these opportunities and scientific journalists write about how productive these meetings are, organised by academics from Harvard University, Stanford University, the University of Zurich and the Max Planck Institute for Human Cognitive and Brain Sciences, just to name a few. As Radin notes, none describes this Nobel laureate as a backward superstitious bumpkin.[10]

On a more practical level, we witness how religious institutions often play an important counterbalance role for people in societies in which the political system has gone wrong. Indeed, they are often the last protection for people when the rest of society's protections fail them, including governments.[11]

While research for some time has suggested that religious people, as individuals, live longer,[12] recent research also suggests religious *institutions* make people 'nicer'. A non-believing scientist, Robert Putnam, Professor of Public Policy at Harvard, and described by London's *Sunday Times* as the most influential academic in the world today, recently released the results of a massive comprehensive survey in the US. Putman says it showed that '... religiously observant Americans are more civic, and in some respects simply "nicer"'.[13] On every measurable scale, religious Americans were shown to be more generous, more altruistic and more involved in civic life than their secular counterparts.

Interestingly, one of the findings was that the content of a person's belief isn't what matters so much as their level of involvement in a religious community. An atheist who comes to church to support a partner will rate as well as any believer on these scores. What can't be denied, according to Putnam and Campbell (co-authors), is that there is something unique about a religious community that has an impact on people for goodness.

As the modern-day philosopher Alain de Botton insightfully writes in his book *Religion for Atheists, A non-believer's guide to the uses of religion*, religion is unique in its capacity to align the bigger picture (or 'the needs of our soul'[14]) to our everyday experiences, as opposed to a secular view of the world, which he suggests 'maintains an irrational devotion to a narrative of improvement, based on a messianic faith in three great drivers of change: science, technology and commerce'.[15]

MEANING

Aligning the specific to the bigger picture is another 'higher-order' ability our neocortex has facilitated for us. Not only do humans represent the capacity of the universe to be aware of itself (for, let's face it, we would appear to be part of the universe), we are arguably the masters of detecting context; that is, the relationship between a specific thing and a bigger picture. It gives purpose to achieving a goal that our values intuitively tell us is desirable, that somehow connects us to a higher, cosmic, reality. For some reason we intuitively sense this as important—well, at least many of us do.

Barney Zwartz writes in *The Sydney Morning Herald* of the rising number of people searching out practices to satisfy their spiritual hunger, which some suggest is the result of today's high-pressure materialistic lifestyles. David Tacey of La Trobe University, a long-time writer about young people and religion, says people are hungrier than ever for the transcendent, an experience beyond themselves; but because there is a fashion against turning to traditional religions, this searching is largely unnoticed.[16] Many people are turning to the East for answers. One of Melbourne's biggest book shops, Readings, finds mystical literature is the most popular after cookbooks and travel guides.

Psychiatrist Louise Newman, director of Monash University Centre for Developmental Psychiatry and Psychology, says values and purpose are part of the human condition: 'We see it today in hugely different ways. The rise of fundamentalism, particularly among the young, is part of this reaction against what we've seen as a decline [in traditional religions]. People are getting on to creeds as a way to re-establish core values.' Newman says that evidence shows that recognition of spirituality and a sense of meaning are fundamentals of mental health, and that ignoring this dimension of our human condition leads to depression, feelings of social alienation, drug and alcohol problems and relationship difficulties.[17]

The abhorrence many of us feel at cases of cruelty towards children, and of torture and injustice, is a further example of our values at work, informing which future we work towards achieving for ourselves, our families and our communities. The fact that some people sacrifice their own lives for their values shows how potent values are. Each day people make non-survival sacrifices for their causes, from a suicide bomber in the Middle East, to a nun shedding her identity and spending her life caring for orphans in an impoverished corner of Africa. These people are detecting a more important dimension to reality than 'survival of the fittest' or the value of self-preservation.

In fact, one could say that the potency of our experience of values provides good testimony that free will exists. Our fulfilment of values appears to be a way of identifying with the greater context of the universe. Values and purpose are irrelevant unless there is free will to define these actions. A determined universe has no ultimate goal, no principles beyond its mechanical chain reaction into oblivion. There is no top-down causation or meaning, as suggested by Bertrand Russell.[18]

This is why some corners of science, believing in a determined universe, have problems acknowledging the agency of values. Some say they don't really exist, that they are a manifestation of our 'survival instinct'. At first glance this appears to be a reasonable position, as one could imagine scenarios that could connect a survival urge to pretty much any behaviour one can observe.

The difficulty is, however, that the survival instinct is considered a *value* in its own right. Moreover, one could use the same process as above to argue that any value is *the* derivative value. For example, 'I kill not to protect myself, but my children ... my people ... my country'—that is, altruism.

Of course, this attracts counter-arguments like, 'Oh, but really he/she does this in order to allow his/her genes to survive'; which is then countered by 'people sacrifice themselves for non-family too; indeed for causes that have nothing to do with the survival of individuals, such as a fireman saving a baby or a surf life saver rescuing someone struggling in the surf', and so on, until the arguments get so convoluted that they lose sense. So this approach appears to lead us to a dead end.

Some have claimed that because electrical stimulation in the brain[19] can simulate the religious experiences and a sense of values, they otherwise may not exist. Again, these same electrical stimulations can also create experiences of colour and light etc,[20] though the ability to simulate the experience doesn't necessarily mean that the electromagnetic wavelengths that normally produce colour are just our fantasies.

In fact, a recent experience I had with a person suffering from a brain tumour provides an example. The tumour was causing her to feel very hungry all the time. Just because the brain can register a feeling doesn't mean that it is reflective of reality. Equally, just because the brain can simulate a feeling doesn't mean that the usual cause of the feeling, that is, the need for the body to replenish nutrients, is illusory.

Religion appears to be society's framework to bring some order and control to these perceptions of ours, and to manage the *burden* of choice regarding which future to bring into reality. One can view it as a burden because in our space-time dimensions there is no going back. Once we've done something, it's done. Our everyday reality has no rewind button. We have no way of really being sure that we are making the 'right' choices.

Religion seems to represent humanity's accumulated experience in managing our perceptions in this regard; painting, to some extent, a picture of what we intuitively sense but can't totally make sense of. It establishes the moral-choice architecture that allows us to manage this responsibility. Most often, it does so in a way that tries to make our common experience a source of unity.

FROM ZERO-SUM TO COMPLEMENTARITY

Modern-day science and religion have undergone a separation in the past two hundred years. And it is probably true that the politics between religion and modern science play a role in keeping both honest. As Sir Nevill Mott, an English physicist who won the Nobel Prize for Physics in 1977, said:

'Science can have a purifying effect on religion, freeing it from beliefs of a pre-scientific age and helping us to a truer conception of God. At the same time, I am far from believing that science will ever give us the answers to all our questions.'

Too often, however, the politics become destructive, like when soccer fans suddenly turn into hooligans. More fascist elements on both sides come out with rhetoric suggesting that the world would be a better place without the other.

More sober minds have found the need to intervene in the game. In 1998 the US National Academy of Sciences released a statement which included the following:

'Religions and science answer different questions about the world ... Religious and scientific ways of knowing have played, and will continue to play, significant roles in human history ... No one way of knowing can provide all of the answers to the questions that humans ask. Consequently, many people, including many scientists, hold strong religious beliefs and simultaneously accept the occurrence of evolution ... Science is a way of knowing about the natural world. It is limited to explaining the natural world through natural causes. Science can say nothing about the supernatural. Whether God exists or not is a question about which science is neutral.'[21]

This statement supports the complementary relationship between religion and science in the pursuit of knowledge, a view also held by some of history's greatest minds, including Albert Einstein,[22] Charles Darwin (irritated by the use of evolution theory by some to support atheism),[23] Sir Isaac Newton (who arguably saw himself as more of a theologian than a physicist)[24] and even Galileo[25] himself. Indeed Gregor Mendel, the founder of modern genetics, was a Christian monk. These scholars, like many others, were aware that religion and science are not opposites, but two sides of knowledge-gathering.

As one Nobel laureate in physics, Max Planck, said,

> 'There can never be any real opposition between religion and science; for one is the complement of the other ... And indeed it was not by accident that the greatest thinkers of all ages were deeply religious souls ... [Science and religion] do not exclude [each] other; rather, they are complementary and mutually interacting. Man needs science as a tool of perception; he needs religion as a guide to action.'[26]

Yet it is worth noticing how fascism can infiltrate modern science, portraying a puritan view of science, ironically based on an all but superseded nineteenth century view of a deterministic universe. Many think the treatment of Rupert Sheldrake by TED, as discussed earlier in this book, is an example of this.

Modern-day discourse allows a place for shock jock 'scientists' to display their intolerance of the religious beliefs of others as a medal of honour. At times they appear to strut about a little like JK Rowling's sneering Lucius Malfoy, obsessed with curbing the influence of what they consider to be the *mudbloods* of the scientific community.

The increasing immediacy of the Internet and news channels has meant that those willing to delve soberly into the complexities and subtleties of the subject are too often passed over for the more dramatic scenes of extremists in battle, regularly portraying the shock jocks as the representatives of the 'scientific view' and conservative religious figures as the representatives of free-will spirituality.[27] While it does make dramatic TV and buoys ratings, one wonders how much these title fights actually do to break down ignorance and prejudice.

RELIGION AND SUBSPACE

What does religion tell us about the existence of subspace?

First, we need to take a look at what religions say. The main religions of today are the product of thousands of years of human contemplation and disciplined thought. Traditional religion is a broad church, with some believing in a 'personal' God[xliv] (Judaism, Christianity, Islam) while others do not (Hinduism, Buddhism and Taoism). This being said, the more one studies, the more obvious it becomes that religions have more in common than not.

Indeed, it is widely noted in the study of mystical teachings, that over the millennia—despite geography, religion, cultural context, or education—these basic teachings seem to be in more agreement on the nature of reality, as Radin puts it, than 'most scientists are able to agree on anything from the causes of climate change to what flavour of ice-cream is best'.[28]

The core similarity of religious principles is a striking characteristic in religious studies. This may suggest, using the triangulation

---

xliv I must say I find it curious how some qualify their disbelief by suggesting that they don't believe in a 'personal' God. I assume by a 'personal God' they are not suggesting their own personal God (like their own personal toothbrush), but one with a personality. This raises the question, why would they give themselves a personality yet deny it in the ultimate creative spirit of the cosmos?

Perhaps they are just suggesting that God doesn't have a long beard and is in the form of an old man. If this is the case, then they are probably right, but I'm not sure how much insight this is adding. Or perhaps they are suggesting that a God exists but has no interest in us or our daily lives.

Again, it is difficult to have any grasp of what a supreme entity like a creator might or might not be interested in, but I confess I would be disappointed if I found out there was a creator and that this being was more interested in the game of cricket than in us.

metaphor, a common *phenomenon* is being perceived, rather than a solely abstract subjective experience.

Although today people are turning to spirituality, they are not necessarily turning to the traditional religions. Tacey warns that there is a danger here: '... it makes people more gullible to cults and sects and various New Age groups who are often asking for people to pay big money'.[29] As British author GK Chesterton suggested a hundred years ago, the danger is that 'when we stop believing in God, we do not then believe in nothing; we believe in anything'.[30]

Assisted by, among other things, a review of acclaimed author and theologian Philip Clayton's summary of different religions,[31] the following is an overview of the major religions today.

Buddhism emphasises the interdependence of all things, as well as the importance of *mindfulness*—presence in, with and to the moment, just as it is and for its own sake. It believes in reincarnation, that our spirit or consciousness is on a journey towards perfection; that is, towards achieving the state of a Buddha.

A Buddhist monk meditating. Buddhism believes in the interdependence of all things. Abbot of Watkungtaphao in Phu Soidao Waterfall.

It believes that there are many worlds and dimensions throughout the universe where spiritual beings like us reside, and that through mindful meditation, compassion and discipline, we are able to make contact with them and gain wisdom about our existence. Similar to modern physics, it proposes that real existence is very different to the world we experience every day, and that what we think is real doesn't really reflect what actually exists.[32]

Compassion and not doing harm to others, including insects, make it a vegetarian religion. Western science does not consider it as much as a threat as other religions, no doubt helped by the popularity of the Dalai Lama, who says he would side with science if ever it clashed with Buddhist tradition.[33]

Hinduism has an ability to recognise the innate value in all individual living things at the same time that it emphasises their grounding in a single ultimate source and reality. It encourages religious practices that are satisfying and effective in their own right, while pointing beyond themselves to a fundamental spiritual orientation to the world.

Hinduism believes the whole world is a single family of religions and that religious practice is healthy in its own right. A statue of Hindu deity Shiva in a temple in Bangalore, India. Deepak Gupta—Meditate.

Hinduism conceives the whole world as a single family that honours the one truth and believes in freedom of worship. It accepts all forms of beliefs and dismisses labels of distinct religions, which would imply a division of identity.

Prominent themes in Hindu beliefs include *dharma* (ethics and duties), *samsāra* (the continuing cycle of birth, life, death and rebirth), *karma* (action and subsequent reaction), *moksha* (liberation from samsara), and the various *yogas* (paths or practices).

Most Hindus believe that the spirit, or soul—the true 'self' of every person—is ultimately indistinct from Brahman, the supreme spirit. Salvation in Hinduism means to be free from rebirths, and to have union with the soul of the supreme spirit.

Taoism is a religion of harmony, balance and co-operation. For Taoist philosophers, whenever one part of society loses sight of its dependence on the other parts and begins to dominate, the balance of Tao is lost, and one should again seek the middle way. They believe in the three basic virtues of compassion, moderation and humility.

Taoism is a religion of balance.

Taoist schools traditionally feature reverence for Laozi, immortals or ancestors, along with a variety of divination and exorcism rituals and practices for achieving ecstasy, longevity or immortality.

Taoism also offers a deep understanding of right action with the concept of *wu wei* ('without action'). Great results can be achieved without effort—*wei we wei* ('action without action' or 'effortless doing')—if one retains the balance of the Tao.

The universe is seen as constantly re-creating itself, as everything that exists is a mere aspect of *qi*, which 'condensed, becomes life; diluted, it is indefinite potential'.[34] Qi is in a perpetual transformation between its condensed and diluted state (intriguingly, similar to the concept of wavelength and particle existence in physics).

These two different states of qi are embodiments of the abstract entities of *yin* and *yang*, complementary extremes that constantly play against and with each other and cannot exist apart.

Human beings are seen as a microcosm of the universe. It is believed that deeper understanding of the universe can be achieved by understanding oneself.

Judaism also believes in an immortal soul, and distinguished itself in ancient times by believing in the one God. Jewish observance draws attention to the importance of consistent practices in the midst of the complexity and ambiguity of human intentions and attitudes. It also places strong emphasis on the importance of justice and the contribution people must make to bringing about justice in the world—not despite their religious convictions but because of them. Jewish thinkers have also emphasised the vast difference between God and human beings; hence the unknowability of God, and the importance of humility about whatever are one's own beliefs and convictions.

Judaism championed the one God. It sees the value of consistent practices in balancing our complex lives and the pursuit of justice. Jewish high priest and Levite in ancient Judah.

Christianity grew out of Judaism and lays great stress on the internal religious life. This religion always asks not only 'What did you do?' but, at the same time, 'What was your inner attitude when you did it?' It places great stress on love. Its founder, Jesus, said to 'love your neighbour as yourself' and that there is a greater game than the trials and tribulations of ordinary life.

Like other religions, Christianity believes each person has a soul, or spiritual form, which survives death. It places great emphasis on belief and faith, where doubt is not to be avoided but a necessary condition for belief or faith to be nurtured.

Christianity believes what you do is important, but second to what you meant to do.

It believes that God's spirit walked the earth as a human, as Jesus—the *only son* of God, not only to provide instruction to humans, but also to experience the challenges of a human and to demonstrate the supremacy of God. Jesus' forgiveness of humans at the crucifixion meant, however, that they were no longer expected to achieve perfection for immortality. Jesus had secured, through his life and death, immortality for humans, past and future.

As a human, it was important that Jesus had doubts like anyone else, particularly about his beliefs and who he was. His supreme act of faith was to sacrifice himself, without any assurance that he was not deluded or insane or just plain wrong (if Jesus had no doubts, there would have been no act of faith). His act reflects what many Christians believe to be an unavoidable choice that comes to each of us at one point in time, which the divine helps prepare us for.

Islam says that devotion to God is the key to our greater life and was one of the first to engage in natural philosophy, which opened the way to modern science.

Islam shares its roots with other Abrahamic[xlv] religions, Judaism and Christianity. It emphasised the sovereignty of God and the absolute responsibility of humans before their maker. The five pillars of Islam set high standards for human devotion to God, reminding the believer constantly that all things come from God and will return to God, so that God really is the final judge and determiner of all that exists and occurs. Yet alongside Islam's high standard for obedience are profound mystical writings of devotion to, and even with, the One and Ultimate.

Islam was one of the first traditions to engage in 'natural philosophy', which opened the way to modern science, and has a particular interest in science. Many Muslims agree with Islamic philosopher Seyyed Hossein Nasr that the Western view of science, which emphasises materialism, naturalism and reductionism, has come to dominate science and is prejudicially hostile to beliefs in God and spirituality.

---

xlv Abrahamic religions (also Abrahamism) are the monotheistic faiths of Middle East origin, emphasising and tracing their common origin to Abraham or recognising a spiritual tradition identified with him.

The *indigenous religions* are the fundamental ecosystems on which later cultures and complex civilisations have been built. They acknowledge the interdependence of humans and animals. Although their relationships were brutal, the interdependence was understood in a way that would often make them thank the animal for its death at a hunt. Many groups had sophisticated theologies centred on natural phenomena, similar to the Abrahamic traditions (Islam, Judaism and Christianity).

Our best guess, and DNA testing, suggests that the Aboriginal culture of Australia is the oldest continuous culture on Earth. They left Africa 75,000 years ago and were the first modern humans to traverse unknown territory in Asia and Australia.[35]

Indigenous religions are the foundation of modern religions. The Aboriginal people of Australia have the oldest continuous culture and believe that the spirit existed before the material body and will after it. Aboriginal rock art, Anbangbang Rock Shelter, Kakadu National Park, Australia. It is said to show Namondjok, a Creation Ancestor, with his wife, Barrginj. Below, the Lightning Man Namarrgon to the right, and a group of men and women with ceremonial headdresses underneath. Source: Thomas Schoch.

The Aboriginal people of Australia believe that the world and their important ancestors were created in the Dreamtime. For them, the Dreaming is meant to describe their various religious beliefs. Indeed, scientists who advocate the holographic view of the universe equate the Dreaming with the 'implicate order' (an 'enfolded' state) put forward by David Bohm (more later).

The Australian Aboriginal people believe in the immortality of the soul or spirit, that it existed before the body and exists after the body in the Dreaming. They believe that animals and plants came into being before humans. Their 'souls' existed and didn't know when they would become physical. When they did become physical, all but one of the souls became plants or animals. The last became human to act as a custodian or guardian to the natural world.

Case studies

Various traditions of yoga are found in Hinduism and Buddhism. Yoga comprises the physical, mental and spiritual practices or disciplines which originated in ancient India, aimed at attaining a state of permanent peace and 'union with the divine'.

The *Yoga Sūtras of Patañjali* are 196 Indian *sūtras* (sayings related to the philosophy of life) that constitute the foundational text of yoga. The yoga sūtras incorporate the teachings of many other Indian philosophical systems prevalent before the Christian era, written by Patañjali more than two thousand years ago, though it is thought that Patañjali incorporated other yoga writings from as far back as 500 years before his time.

The *Yoga Sūtras* are built on a foundation of sāṃkhya philosophy and are considered to be the practice coming out of sāṃkhya theory. The sūtras diverge from early sāṃkhya by the addition of the principle of *Isvara*, or God, and that surrender to God is one way to liberation. Another divergence from sāṃkhya is that while the sāṃkhya holds that knowledge is the means to liberation, Patañjali's yoga emphasises the methods of concentration and active striving. The aim of yoga is to free the individual from the clutches of matter, and it considers intellectual knowledge alone to be inadequate for the purpose.

In his recent book *Supernormal: Science, Yoga, and the Evidence for Extraordinary Psychic Abilities*, Dean Radin investigates the *siddhis*, the techniques that are known to achieve what people in the West consider supernormal abilities within yoga. These techniques include, for example, ecstatic dancing, drumming, praying, chanting, sexual practice, fasting or ingesting psychedelic plants and mushrooms.

'The yoga sūtras tell, in a matter-of-fact way, of the exceptional abilities that will come to those who sit quietly, pay close attention to mind and practise diligently.' Dean Radin.

The sūtras tell, in a matter-of-fact way, of the exceptional abilities that will come to those who sit quietly, pay close attention to mind and practise diligently. They are not considered *magical* or *exceptional*, just by-products of the road to enlightenment. Siddhis may include experiences of telepathy, clairvoyance, precognition and psychokinesis. More advanced forms include invisibility, levitation and super strength.

It is suggested that these abilities come with the mind's realisation through disciplined meditation that we are all a part of the same identity, when the distinction between the observer and the observed dissolves. This paves the way to enlightenment.

The siddhis include the ability to simultaneously perceive the past, present and future. While our memory shows us how the past is attached to the present, there is now experimental evidence in today's physics, as discussed above,[36] that the present is constrained by the *future* as well as the past. As Radin states, the originators of this concept aren't mystics but include such scholars as physicist Yakir Aharonov, awarded the US National Medal of Science in 2010 and regarded today as one of the world's leading quantum theorists.[37]

As Radin writes, 'tales of supernormal mental power are not unique to the yogic tradition. Most of the same abilities are described in Catholicism as *charisms* and in Islam as *karamats*. In Judaism, *nahash* or divination may be practised by a *tzadik*. In Tibetan Buddhism, the term is *ngon she*, meaning heightened awareness ... all shamanistic traditions are saturated with such tales.'[38]

Patañjali warns against getting attached to these abilities when they manifest and writes about them to prepare practitioners for their arrival, so that they do not overreact and allow themselves to be distracted by them. Boasting about these abilities is also frowned upon in yoga tradition and considered a distraction from enlightenment.

Radin makes the point that while these abilities are considered exceptional in the West, in the larger part of the world, such as in India, they are considered a regular part of the everyday. The fact that Western science has not caught up with these phenomena does not appear to be of relevance or concern to them. In fact, the concept that the mind is not reducible to the brain is not considered *cutting edge* or *profound* or *fringe*; it has been experienced by yoga and many other traditions through thousands of years of first-hand experience and intelligent contemplation.[39]

In his book, Radin provides examples of research showing how yogis who are practised in the siddhis can demonstrate exceptional abilities. He tells of people able to control their body temperature to allow them to survive what would be otherwise deadly cold temperatures. This ability is demonstrated by Tibetan monks and has been confirmed by Harvard University's Herbert Benson and colleagues.[40]

He examines studies of *breatharianism*, or what is referred to in Catholicism as *inedia*, the claim of living well without eating food. He provides examples of tested cases, such as the Indian *sādhu* Prahlad Jani. He claims he was visited by divinity as a child and was told that he would not have to eat food. He is said not to have eaten anything for most of his eighty-one years.

Jani was tested in 2003 and again in 2010 at Sterling Hospital in the western Indian city of Ahmedabad, by neurologist Sudhir Shah and his team. In the 2003 test, Jani was monitored around the clock for ten days by hospital staff and video cameras. The results of the test seemed to confirm Jani's claim, with no signs of food or drink being consumed or any change in his physiological condition. Not even going to the bathroom. In 2010 he was observed by a team of 35 researchers from the Indian Defence Institute of Physiology and Allied Sciences, and other organisations. Again the claims appear to be confirmed.[41]

While discussion ensues about the rigor of these experiments, Radin presents an encyclopaedia of research demonstrating psi abilities within the rigor of the laboratory, such as those described above in the Biological section, comparing meditators' (including yoga practitioners) abilities with those of non-meditators. Radin's work shows convincing evidence in support of the psi-nurturing properties of yoga practice.

He is not alone. Russell Targ also claimed in a recent interview that after 20 years of psi research for organisations such as NASA and the CIA, *'what we discovered is that the ancient Hindu Patañjali ... had it right...'*[42]

In Christianity, deep meditative practices have produced similar phenomena. One case in particular involved a young Bavarian woman by the name of Therese Neumann.

She was born on Good Friday, April 8, 1898, in the small village of Konnersreuth, and was the eldest of ten children. She aspired to become a missionary in Africa but after a series of accidents as a young adult, Neumann was eventually left paralysed, bedridden and totally blind in 1919. She suffered bedsores constantly. Dr Mittendorfer, a Munich physician who attended Neumann, stated that these sores were so deep that her bones were exposed.

She found comfort in spiritual readings, so her family members and Father Naber, pastor of the parish of St Lawrence in Konnersreuth, read to her aloud about the lives of Jesus and the saints.

After four years of suffering paralysis and blindness and a series of Catholic events which had meaning for Neumann—including the beatification ceremonies in Rome of the Carmelite nun Thérèse of Lisieux (a favourite of Neumann's), she was completely healed. In 1926 Neumann reported a series of semi-conscious episodes and visions of the events surrounding Jesus' crucifixion. Shortly afterwards, she exhibited *stigmata*.

Therese Neumann was born on Good Friday, April 8, 1898, in the small village of Konnersreuth, Bavaria, Germany, and exhibited miraculous abilities, including stigmata and inedia. Source Ferdinand Neumann, German Federal Archives.

Stigmata (singular *stigma*) is a term used by Christians to describe, in profound cases, wounds appearing on the body corresponding to the crucifixion wounds of Jesus Christ, such as on the hands, wrists and feet. St Francis of Assisi was the first recorded stigmatic. Many cases followed. Neumann's was particularly profound. She experienced wounds above her heart,[xlvi] which for some time she kept secret, and then wounds to her hands and feet (as if pierced by nails). Later, after she reported receiving more visions of the crucifixion, she even exhibited wounds to her head as if a result of the crown of thorns. At times she also bled from the eyes.

Her case was one of the most investigated and documented. Her wounds, like those of St Francis and many other recorded stigmatics, never got infected and generated protuberances from the wounds resembling nails. They appeared as hardened skin and under the examination of several medical doctors were found to be structures that passed completely through her hands and feet. Unlike the wounds of St Francis, however, Neumann's opened only periodically, and when they stopped bleeding, a soft, membrane-like tissue quickly grew over them.[43]

Investigations—including those done by Herbert Thurston, an English priest who documented many cases of suspected miracles, and

---

[xlvi] Represents the place where the lance of Longinus penetrated the sacred body of Jesus.

author of *The Physical Phenomena of Mysticism*—concluded that stigmata were more likely a product of autosuggestion than a divine reenactment of the crucifixion of Jesus. A variety of factors are put forward to support this view, including the observation that stigmata vary from person to person, that a large number of stigmatics experience hysteria (however, this is no longer considered a diagnosis) and that the wounds appear through the palms of hands and resemble artists' impressions of the crucifixion; this differs from historical evidence, which suggests the Romans drove crucifixion nails through the wrist.

Other abilities demonstrated by Neumann, which appear to be psychokinetic in nature, seem to support this assessment. Witnesses reported that Neumann's foot wounds would bleed towards her toes, even when her feet were pointing upwards. In other words, her blood would flow up, against gravity.

However, perhaps even more remarkable was her well investigated and tested inedia. This began when she transferred a throat disease from a young priest to herself. She endured the disease for several years, while consuming only liquids; that is, she consumed no solid food. Then, in 1927, Neumann gave up food and water entirely.

Neumann's claim, of course, attracted considerable scrutiny; for example, the local bishop commissioned a medical doctor to supervise four Franciscan nursing sisters to watch Neumann day and night for fourteen days. The directions to the sisters were very strict and carefully calculated. Neumann was not to be left alone for a single moment, whether at home, in church or out of doors, day or night. Even her customary confession was forgone. The sisters were to bathe her, but with a damp cloth instead of a sponge. The water for mouthwashes was to be measured and remeasured to guard against it being swallowed. She was periodically weighed. Her pulse was recorded, temperature taken and blood samples collected and analysed at prescribed times. According to the *Münchner medizinische Wochenschrift* supplement No. 46, 1927, the directions went so far as to demand that 'All excretions—urine, vomit, and stools—must be gathered, measured and weighed, and immediately sent to the physician for analysis.'[44]

During the testing it was found that Neumann never went to the bathroom. When she lost weight during her blood loss at stigmata periods she regained it quickly afterwards. She suffered no ill effects and showed no signs of dehydration even at the fourteen-day mark. The doctor who supervised the test gave evidence later in court that Neumann had not eaten or drunk a thing during the examination.

Professor Ewald of Erlangen, a sceptic at the time, admitted that the keenest and most relentless attention was given to the matter of food throughout the period of observation. He claimed that Neumann ought to have lost weight heavily, but such was not the case. In relation to the loss of liquids due to normal breathing, bleeding and perspiration etc., 'Therese ought long since to have been dried up like a mummy. But she is fresh-looking and lively, has saliva, and moist mucous membrane. One may indulge in the most fantastic imaginings, a prolongation of metabolism as in hibernation, or *iakirism*—though Therese does not hibernate, but moves, speaks, reads, writes letters, goes about—this increase in weight simply cannot be explained; for something cannot come from nothing.'[45]

Neumann reportedly did not eat or drink for the rest of her life, and died in 1962. If the records are accurate, it would appear that Neumann exhibited her mind's ability to materialise the liquid, nourishment and/or blood she required.

The different strands of Buddhism are very much aware of the profound effect of meditative states. The Dalai Lama speaks of the 'miracles' of the mind and the implications for the nature of existence. He writes that mind 'enjoys a status separate from the material world ... from the Buddhist perspective, the mental realm cannot be reduced to the world of matter, though it may depend upon that world to function'.[46] One could easily mistake this statement as coming from a quantum physicist.

Max Planck, for example, the originator of quantum theory, was not as diplomatic as the Dalai Lama, and once said in an interview, 'I regard consciousness as fundamental. I regard matter as derivative from consciousness. We cannot get behind consciousness. Everything that we talk about, everything that we regard as existing, postulates consciousness.'[47] As Niels Bohr, a Danish physicist who made significant contributions to understanding atomic structure and quantum theory, once said, 'everything we call real is made of things that cannot be regarded as real'.

Today theoretical quantum physicist and author Amit Goswami, former professor and teacher at the University of Oregon's Department of Physics for thirty years, is among a growing number of scientists investigating consciousness and spirituality to shed light on discoveries in modern physics. For Goswami, consciousness, not matter, is the ground of all existence. 'If ordinary people really knew that consciousness and not matter is the link that connects us with each other and the world, then their views about war and peace, environmental pollution, social justice, religious values and all other human endeavours would change radically,' says Goswami.[48]

In his sessions with Western scientists, the Dalai Lama has said, 'there are instances where small children recollect their past life very vividly'. He talks about Tibetans who dreamt about the path across the Himalayas, and years later, when they actually had to cross the Himalayas, they felt familiar with the route. He also tells us that 'there are accounts of people experiencing the sense of leaving their body, actually perceiving things in the external world, and later being able to recall events that presumably took place there, even to the point of being able to read a book in someone else's home'.[49] He asks whether science has ever investigated these things. Unfortunately, he asked scientists unfamiliar with the above studies at the time.

How it must look to ancient cultures, watching some in Western science strut about suggesting a phenomenon doesn't exists until Western science discovers it!

DISCOVERING THE SCIENCE BEHIND RELIGION

Émile Durkheim lived in the late nineteenth and early twentieth centuries. He was a French sociologist, social psychologist and philosopher and is widely considered (along with Karl Marx and Max Weber) the principal architect of modern social science and indeed the father of sociology.

He advocated the study of religion scientifically and considered what he termed *collective effervescence* to be the cause of religious vitality. He saw collective effervescence as a phenomenon which transformed a group of individuals into a single group-consciousness able to generate social and religious change. He was inspired by studies of the Australian Aboriginal Arunta people and reports of their ceremonial rituals, but also saw the phenomenon in the Crusades and in the French Revolution where *effervescence* was directed towards French nationalism.

Émile Durkheim, known as the father of sociology, wanted science to study the nature of the 'other world' people go to when enter a group consciousness—Public Domain.

He saw collective effervescence as a 'mobilisation of all our active forces, and even a supply of external energies'. He said that once a person arrived at this state of exaltation with others, they no longer recognised themselves as individuals, but were 'carried away by some sort of an external power which makes [them] think and act differently than at normal times ... as though [they] really were transported into a special world'.[50]

Unlike others of his time, he did not see this phenomenon as abnormal, or even an epiphenomenon (a secondary phenomenon caused by another). Nor did he see it as a mental illness or some other sort of pathological event or state, as some psychologists interpreted it. As Arthur Buehler, senior lecturer in religious studies at Victoria University, points out, Durkheim considered it an experience of unitary being, a real and possibly a *post-rational* state.[51]

Furthermore, Durkheim was convinced that the phenomenon could be measured scientifically. But as Buehler also points out: since Durkheim, hardly any anthropologists have taken Durkheim's challenge to measure or observe collective altered states seriously or gathered data to confirm or refute collective effervescence.[52]

For Buehler and others, this is due to a number of factors. First, it is a remnant of colonial sentiments that considered non-white spirituality as primitive and made up of the illusions of ignorant or 'lesser' people.

Second, he sees the reluctance to accept Durkheim's challenge as an example of *armchair scholarship*, the tendency for some scholars—particularly in Durkheim's time (including Durkheim himself)—to study cultures without actually visiting and experiencing them, often resulting in ethnographic material being used merely to support pre-existing ideas.

Third, Buehler sees it as a result of a prevailing scientific materialist paradigm that was prevalent in Durkheim's time and is still strong today, a view that sees the universe arising solely out of physical events originating from the Big Bang, as discussed in previous chapters. This ideology considers mental processes—consciousness, including spiritual or religious sentiments—as, at best, epiphenomena, or even a pathology (attributing mental illness in those who claim to experience altered states of consciousness).

The materialist view places a taboo on subjectivity, and analyses social phenomena solely through a lens of 'rational' consciousness, frowning upon anthropologists *going native* by experiencing at first hand the subjective reality of the cultures and ceremonies which they study.

Buehler explores how these views have changed with the increased sophistication of anthropology over the past hundred years. Contemporary approaches at times require researchers to experience the subjective reality of those being studied, and even to partake in ceremonies and rituals, using research methods that change the investigator's own state of consciousness.

He compares the refusal of some to consider the realness and value of other conscious states as similar to the refusal of scientists to accept Galileo's invitation to look through his telescope. Indeed, Galileo feared not the Jesuits, whom he considered 'friends of science and discovery', but the professors at the university.[53]

Durkeim's views do appear before their time. In taking them forward, scholars are accepting that there are other states of consciousness that reflect reality, beyond the normal awake consciousness which much of Western science still chains itself to. In this respect, Western culture is in the minority, as most of the world's cultures value multiple states of consciousness.

Buehler and others argue that we can honour both objective data and data from our subjective and inter-subjective experiences, and that this is, indeed, the expectation of twenty-first century anthropologists.[54]

The foundations of sociology certainly do make room for the suggestion that at the base of the religious experience is a mental communion with another world, or even another dimension of reality. It is also likely that we have neglected the serious scientific study of these states and their relationship with society, states which many of us intuitively sense as connecting us with ultra-realities.

We have accomplished astonishing technological feats by getting behind our intuition in the mechanical world. Yet despite this, humans are still asking themselves the same contextual questions today, as they fly from one continent to another in satellite-guided jumbo jets, as they did a hundred thousand years ago under the stars around a campfire: why are we here? Where did we come from? Is there divinity? Does our sense of self survive our physical bodies?

It seems we are yet to get behind our intuitive sense in this area, as we already have in other areas of science. In this respect, science may have let us down by neglecting issues which are arguably more central to our humanity.

Many seem to have given up on a common answer, retreating to their own portion of the *elephant*, and have accepted that—for the moment, at least—either the *elephant* doesn't exist, or it is not worth knowing. This has helped not only divide religious knowledge, but separate religion from other forms of knowledge-gathering.

Why? More of us are asking this question, and not just theologians, philosophers or sociologists, but the brightest minds in the physical sciences facing discoveries that take them to the limits of their own specialty, as we have discussed above.

We have discussed above the views of Buehler and others about the reasons for this, but many also see it as a result of allowing knowledge to grow in silos; that is, disciplines pursuing a narrower and narrower path of specialisation, in competition with other disciplines for a shrinking pool of research resources. Many see it as having the effect of making science more partisan; in turn leading to the loss or neglect of important connections.

We see the symptoms of this: such as specialised disciplines overstepping their limits, trying to sell answers to questions obviously bigger than they are. Take the spotting of the Higgs boson,[xlvii] for example, what some particle physicists have described as the *God particle*.[55] Finding the Higgs boson has been a major goal for the £10 billion Large Hadron Collider (LHC) operated by the European Organisation for Nuclear Research (CERN). A less-powerful machine had failed to find the missing particle before it closed in 2000. Some particle scientists suggest that its discovery will tell us why we exist.[56]

Unfortunately for the LHC, and despite the best efforts of particle physicists, the close encounters with the Higgs failed to create a stir in the religious community and had trouble attracting as much interest in the secular community as the launch of a new iPhone.

This is not to say that particle physics is a waste of time. Of course it isn't. It is just that you can't answer a *why* question with a *how* answer, as illustrated by Douglas Adams in his book, *The Hitchhiker's Guide to the Galaxy*, when a computer proposes that the answer to life, the universe and everything is '42'. It is like explaining to someone that their bus was late because of the elapsed time between when the bus was expected and when it arrived. Our neocortex makes humans pretty savvy customers when it comes to recognising inappropriate answers

---

xlvii The Higgs boson, or Higgs particle, is an elementary particle initially theorised in 1964 whose discovery was announced by CERN on 4 July 2012. The discovery has been called 'monumental', because it appears to confirm the existence of the Higgs field. It would explain why some fundamental particles have mass, and why the weak force has a much shorter range than the electromagnetic force.

to higher-order questions, despite the best efforts of some overenthusiastic particle physicists.

As can be seen with the Higgs boson, competition for scarce resources will entice specialty areas to overstretch their area of expertise, but the outcome does little credit to the scientists or science in general. Nor does it satisfy the general public: it either reaffirms for them the inability of science to shed light on our important mysteries or, worse, that the answers are beyond the reach of mere mortals who don't have a PhD in particle physics.

I wonder if there'll come a day when people experiencing an existential crisis will approach a physicist for solace, instead of a priest or counsellor. For example, some suggest that knowing how big the universe is, and how insignificant we are as a consequence, can provide the solace people need in such times.

As Alain de Botton writes, we would do well to meditate daily not on God, but 'on the 9.5 trillion kilometres which comprise a single light year ... perhaps, after the main news bulletin and before the celebrity quiz, we might observe a moment of silence in order to contemplate the 200 to 400 billion stars in our galaxy, the 100 billion galaxies and the three septillion stars in the universe ... We would then be able to ensure that our frustrations, our broken hearts, our hatred of those who haven't called us and our regrets over opportunities that have passed us by would continuously be rubbed up against, and salved by, images of galaxies such as Messier 101, a spiral structure which sits towards the bottom left corner of the constellation Ursa Major, 23 million light years away, majestically unaware of everything we are and consolingly unaffected by all that tears us apart.'[57]

While I doubt that Botton was suggesting that these contemplations alone were sufficient to replace modern religion, one wonders about the ability of numbers to meet our existential needs, even if they are significantly greater than 42 (*The Hitchhiker's Guide to the Galaxy*).

POSSIBILITY OF A SINGLE CONSCIOUSNESS

The possible existence of a single consciousness is a popular idea, which reverberates in religious as well as modern scientific discourse—as expressed in the quantum soul theory,[58] as put forward by Stuart Hameroff and Sir Roger Penrose.

We know from quantum physics that it is possible for a particle of light to travel to us from a galaxy millions of light years away in a quantum potential state (see above discussion)[59] until the moment it reaches us (the conscious observer). At this point it becomes an event—and enters our reality as a material object (particle).[60] Its history as a particle is created at that point, going back millions of years. As Wheeler proposed, this could mean the universe is brought into reality retrospectively, through observation,[61] a view shared by Hawking.

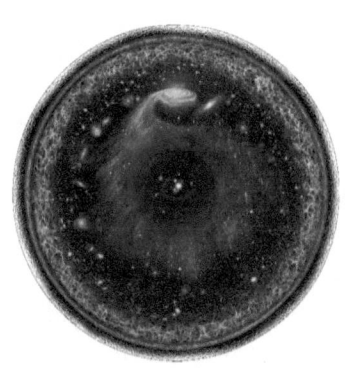

'Is it possible that consciousness, like space-time, has its own intrinsic degrees of freedom, and that neglecting these will lead to a description of the universe that is fundamentally incomplete?' Stanford University physicist, Andrei Linde. Graphic: Pablo Carlos Budassi. Effect added.

Wheeler suggests that the universe may have come into being with the first observer, collapsing the potentiality of all possible universes (for example, all possible universes in superposition) into our actual universe—the only universe able to sustain our material existence/form. In this scenario the 'first cause' is an immaterial consciousness, with information. This consciousness created the universe, as it was the only one capable of supporting our material form. Perhaps Max Planck was right when he said, 'I regard consciousness as fundamental. I regard matter as derivative from consciousness.'[62]

We know that the moon, like each of us and all matter, is mostly 'ghostly empty space'. It is the emergent property of non-physical phenomena *in-formation*. Our perception of it as hard matter etc. is only how our minds interpret and make sense of the information. Its true nature is beyond our current understanding.

Einstein pondered whether the moon was there if we were not looking at it. This question was echoed in modern times with confirmation of the observer's role in bringing matter into the physical world, that we are presented with the possibility that the moon's existence in material form is conditional on our (or a) conscious perception of it.

As there is every reason to believe that the moon is the same for you as it is for me—that when I look up, I am seeing the same information that you are—it appears theoretically possible that a consciousness exists maintaining the moon as it is, while neither you or I are looking at it.

Stanford University physicist, Andrei Linde speculates, 'is it possible that consciousness, like space-time, has its own intrinsic degrees of freedom, and that neglecting these will lead to a description of the universe that is fundamentally incomplete? What if our perceptions are as real (or maybe, in a certain sense, are even more real) than material objects? What if my red, my blue, my pain are really existing objects, not merely reflections of the really existing material world? Is it possible to introduce a "space of elements of consciousness", and investigate a possibility that consciousness may exist by itself, even in the absence of matter, just like gravitational waves, excitations of space, may exist in the absence of protons and electrons?'[63]

One wonders how the Dalai Lama and other religious leaders would respond to this question.

SUMMARY—TRIANGULATING RELIGIONS

Our neocortex has given us a profound ability to perceive meaning and context, an ability to sense the signs of a bigger cosmic picture. It also picks up universal values that guide us in exercising the apparent capacity of our consciousness to interfere with the chain reaction of space-time mechanics, and determine which future to bring into existence.

Religion appears to be the framework we use to support us in managing the responsibility, the result of millennia of human experience and accumulated wisdom. Our studies show that rather than being in opposition to science, religion is in fact a sponsor of it, a fact understood by some of the greatest scientific minds in history. Indeed, it has been instrumental in creating the civilisation we experience today and underrated as a source of practical knowledge.

Religion appears to present common views of reality. For example, of a spiritual soul that exists along with our material body, the existence of a realm beyond our material everyday world, and our mission to build a loving and responsible personal spirit.

The fact that it provides similar messages across time, geography and culture supports the view that it is indeed a reflection of an ultra-reality, or a subspace-type realm, that provides us (our consciousness) with information that spans time and space. In so doing it links our actions, or free-will activities, to a cosmic whole. Our need to act in ways that unify us with these values reveals a deep need to connect with this higher cosmic existence.

One of the findings, such as those coming out of research into yoga, confirms that religious practice can nurture abilities that traverse the barriers of space-time and access subspace. Durkheim and others also called for the study of the phenomenon he described as collective effervescence, the ability of humans to join together into a single consciousness, and which lies at the core of religious vitality.

Yet, while we have built practical knowledge out of our intuitions in mechanics, we have not been as successful in getting behind the science of religion. There are many possible reasons for this, such as cultural disdain or just the tendency of the phenomenon to rest between silos in a fragmented scientific marketplace.

It might also be because of our need to preserve something that we consider is personal, something we don't need to negotiate with the rest of society or its institutions.

However, as Botton concludes in his book, *Religion for Atheists*, 'the wisdom of the faiths belongs to all of mankind, even the most rational among us, and deserves to be selectively reabsorbed by the supernatural's greatest enemies. Religions are intermittently too useful, effective and intelligent to be abandoned to the religious alone.'⁶⁴

CHAPTER 5
Finding the elephant

Imagination is as important as knowledge.[xlviii]

Accessing a subspace realm within which time and space do not present the same kinds of barriers would propel the human race into a whole new age of possibilities, in space travel, health and knowledge. Is it possible that our specialisation and partisan approach to knowledge-gathering has blinded us to phenomena like subspace, too broad in nature to be understood or captured by one discipline alone? If the research we have reviewed is anything to go by, the answer is likely to be yes.

---

xlviii I provide this comment because, according to the Hebrew University in Jerusalem, which holds copyright to Albert Einstein's material, there is doubt as to whether Einstein actually said 'I am enough of the artist to draw freely upon my imagination. Imagination is more important than knowledge. Knowledge is limited. Imagination encircles the world.' As quoted in *What Life Means to Einstein, an Interview by George Sylvester Viereck*, in *The Evening Post*, 26 October 26, 1929. http://www.saturdayeveningpost.com/wp-content/uploads/satevepost/what_life_means_to_einstein.pdf. What is clear from Einstein's commentary is that he was very much against the approach of some schools and other institutions to education at the time, content merely with filling students' heads with facts and figures and other bland information. He was concerned that it did not provide the creative aspect of education, which challenged and applied imagination to facts. Indeed, Einstein is probably known as the most creative genius in the sciences. For this reason I provide my quote above, as I think Einstein may agree that while knowledge is important, imagination is no less so. However, perhaps the more interesting thing about the quote being often attributed to Einstein is the fact that people generally seem quite prepared to accept that one of the greatest scientific geniuses who had ever lived said it. Perhaps at some deep level, we all think it is likely to be true.

Photo from article from where Einstein's imagination quote originates. Dictating to his secretary. The Evening Post, October 26, 1929.

Despite the enthusiasm of nineteenth century naturalism, the possibility of a top-down as well as a bottom-up reality was never truly dismissed by serious science. The irony is that it was the naturalist approach which ultimately led to advances in quantum physics, the field that turned the naturalistic understanding of reality upside down. The sanctity of the physical laws, the absolute nature of space and time (often termed *locality*), and the (so thought) impotence of consciousness all fell like dominoes as a result of what is now widely recognised as the most tested area in science, quantum mechanics and Einstein's theories of relativity.

THE DIFFERENT PARTS

What now do we find when we bring *the elephant* together?

Quantum research tells us that the material world is all but an illusion. It shows that space (distance) and time are also phenomena manufactured out of a more primal subspace dimension.

The quantum world reveals a tell-tale seam in existence, where dimensions intersect. We witness information move back and forward in time, and witness evidence of particles connectedness across time and space. Our everyday space-time reality does not account for this phenomenon.

A greater mystery is our ability to share an experience of now, to perceive a clump of time at once, which travels in one direction of causality, that is, from the past into the future. According to physicists, our everyday space-time reality does not account for this type of phenomenon.

We have also discussed the role of consciousness in establishing the context of knowledge; that is, a perspective over space-time, allowing us to observe and predict, without being caught up in the chain reaction of the space-time universe.

Physics has also become aware of other tell-tale seams in the cosmos. None more profound than the relationship the mind has with bringing matter, and perhaps our reality, into existence out of a sea of possibilities. It points to the role of consciousness in maintaining the physical or material reality we experience.

We have also observed the agency (action or intervention producing a particular effect) of consciousness, seemingly working from outside of space-time, such as the placebo effect and research into telekinesis (or psychokinesis). This includes the power of meditators to interfere with particles. The effect of the mind upon matter is as established as anything can be in science. While it may not at this stage include lifting X-wing fighters from an alien swamp, as Lucas's character Yoda did in *Star Wars*, the hard part has been overcome, that is, establishing that the effect is real. The rest now is 'merely' an exercise in amplification, something we might leave to rock concert 'roadies'.[xlix] Their time in the limelight may have finally come.

We see our ability to perceive values, to tune in to principles that connect us with a dimension outside of space-time. They provide us with the important navigational tools in managing the burden of choice, the responsibility that comes with being able to choose which future, out of a sea of possibilities, we bring into reality. Our ability to 'see' these values is so strong, it is likely to explain our human behaviour better than any other factor. It is the foundation of religion and

---

xlix 'Roadies' are the road crew, the technicians or support personnel, who travel with a band on tour, and handle every part of the concert productions except actually performing the music with the musicians. This catch-all term covers many people from tour managers to lighting and sound technicians, including those responsible for amplification equipment. Some consider them the unsung heroes of the music industry.

thus civilisation, and it is enshrined in our legal systems, which equate a person's inability to perceive values with insanity.

Tragically, today, as the world grapples with organisations like the Islamic State, the consequences of underestimating the importance of values and meaning to the development of young people is becoming all too clear.

TRIANGULATING THE PIECES

How does triangulation put this together? From what we have reviewed, we can assume that putting the puzzle together is an exercise that requires insight, a *right-brain* function, as much as analysis (*left brain*). It finds connections that we don't already have and which are likely, if Penrose and Hameroff are correct, to be the function in our brain that allows us to use quantum mechanics to identify the idea in the cosmos most relevant to the parts we have assembled.

David Foster Wallace, a highly acclaimed American writer, once told the story of two young fish swimming along. They happen to meet an older fish swimming the other way, who nods at them and says, 'Morning, boys. How's the water?' And the two young fish swim on for a bit, and then eventually one of them looks over at the other and says, 'What the hell is water?'

In telling the story, Wallace meant 'that the most obvious, ubiquitous, important realities are often the ones that are the hardest to see and talk about.' It is quite possible that subspace is one of these realities. It is the 'elephant in the room' that permeates our environment. We are immersed in it, and nothing functions the way we know it does without it. We are like the fish in the water, taking it for granted so much that we fail to see its significance and importance in defining our reality.

When we consider that we see with our brains and not our eyes, this is kind of understandable. Two-thirds of our brain are involved in processing visual sensory input received through the eyes. This is why visual problems are symptomatic of the brain and spinal-cord disorder multiple sclerosis.[2] As discussed in the first chapter, despite a popular misconception, we do not see a direct representation of external reality but a translation formed by our eyes and mind. The study of visual illusions and how they work shows us that reality and human perception of reality are different things.[3]

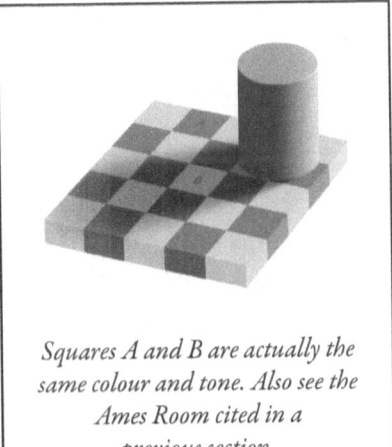

Squares A and B are actually the same colour and tone. Also see the Ames Room cited in a previous section.

Visual information coming from our eyes is processed, edited and modified by our temporal lobe before it is passed to our visual cortices, where our mind 'sees' or perceives it.[4]

Only about half of what the eyes actually see gets through the filtering, and this makes up only half of what our minds see. The rest of the picture is put together by the brain, based on what we expect to see as a result of a whole range of factors, including our culture, beliefs and past experiences. Michael Talbot, author of *The Holographic Universe* (more later), suggests, 'this leads us to a disturbing question. If we are seeing less than half of what is out there, what is out there that we are not seeing?'[5] Indeed, our inability to see some things may be an evolutionary mechanism; that is, at some stage in our genetic history more of us survived by *not* seeing something in our environment that our eyes physiologically perceived.

Could we physiologically have some of the remarkable sight abilities of animals, such as the ability of birds to see Earth's magnetic field, the ability of sharks to see electromagnetic fields or the ability of snakes to see heat? If our brains were rewired, or tweaked here and there, could we even see into subspace? Perhaps, as discussed in the above chapters, some already do.

Is subspace a mega-phenomenon? It appears likely. It becomes visible for what it is, at least on an intellectual level, when we bring together different areas of knowledge, particularly those areas that politics have kept apart over the last few hundred years. As you might appreciate from the above, once you start seeing it, it is nearly impossible to stop seeing it. It reveals itself everywhere, from physics to religion to our legal system. Yet even then it is difficult to understand, and we need to allow room for the possibility we may never fully understand it. In fact, we may have taken only the first step in a thousand-year journey to its discovery.

Perhaps we can start to fathom it by considering its potential to add to our lives in practical ways, once we accept its existence and start working with it. For me, I observe that the immense power unleashed by a nuclear bomb starts off with phenomena at a microscopic level. For scientists, the key to subspace may be in understanding how our minds can initiate a similar chain reaction, but instead of creating an explosion, it would be used, for example, to bring something into, and possibly through, subspace, perhaps to another time and/or destination.

A lot of things start off very small. Our minds might be able to achieve changes at the micro level, and there is no reason that these changes can't also result in powerful changes at the macro level. Atomic blasts start from a chain reaction at the atomic level. Underwater atomic test 'Baker', Bikini Atoll, Pacific Ocean, 1946.

'Where No One Has Gone Before' is an episode of *Star Trek: the Next Generation*. In this episode, The Traveller, whose real name is supposedly unpronounceable by humans, is a mysterious humanoid native to Tau Alpha C, who has the ability to alter space, time and warp fields with the power of his mind. He can phase out of time and dimension and move between planets and starships. He can also shape-shift to appear as humanoids from other species with different clothing.

These abilities were based on his advanced understanding of reality and his ability to focus the energy of thoughts. According to The Traveller, thought was the basis of all reality ('Where No One Has Gone Before', 'Journey's End').[6]

Could our future look like this, with human minds and machines coming together to harness subspace or, as physicist David Bohm might call it, the *implicate order*?

How do we navigate a dimension so foreign to our understanding, one which appears to connect everything, everywhere through time, where mental phenomena are more real than matter? The answer may lie in one of our neocortex's other gifts, our unique pattern-recognition skills, enabling us to perceive higher-order contextual information.

Our pattern-recognition skills have provided us with a profound ability to perceive meaning and context, an ability to sense the signs of a bigger cosmic picture. It is the ability to perceive values that guides us in exercising a capacity to choose which future to bring into existence. In fact, it changes what would be a determined mechanical event into a *decision* event, bringing with it a sense of responsibility.

Our religions could be our institutional attempt to provide ourselves with a framework to manage this responsibility, an idea supported by some of the greatest scientific minds in history. The mental discipline which this framework promotes may hold vital clues as to how we can navigate the mental ultra-reality of subspace.

At a basic level, it suggests that *how* we do something is as important as *what* we do. The former brings us in touch with our existence in a more cosmic reality, visible from subspace. In the near future we may better understand how values like compassion are not just a fashion of human life, but a cosmic principle that is as real and potent as any of the natural forces in space-time.

This raises the prospect of a dynamic link between material mechanics, consciousness and meaning, and points the way to the possibility of a collective consciousness (an extension of the global consciousness phenomenon discussed in a previous chapter) that has a role in determining what advances are made and when.

Have you ever had the experience of trying to solve a puzzle, only to be stumped at every turn? Upon finding out the solution, you think to yourself that the answer was so obvious you couldn't believe you missed it for so long. Could it be that various levels of consciousness have a role in mediating such breakthroughs? This might seem far-fetched, but the research cited above conducted by Sheldrake and others could be interpreted as supporting this theory—when a puzzle is solved in one place, suddenly it is solved more easily all around the world.

When considered alongside Penrose's observations of the fine-tuning of mathematics and the general fine-tuning of the universe (anthropic principle), one might theorise that consciousness has a role in enabling or *allowing* such breakthroughs. Jahn and Dunne of the PEARs lab consider that reality could be created to suit what we want to find. Discovery could be as much creation as finding.

IDEAS AS EMERGENT PROPERTIES

One of the lessons we could learn from this is the importance of bringing the pieces together. We need to question and rationalise the separation modern society likes to use to order things. Our unravelling of subspace might not occur until the conditions required to do it, mentally and physically, come into place. Keeping complementary areas of knowledge artificially apart may prevent this from happening.

Plato thought that ideas were more like solid objects, which had their own existence. A suspicion, by the way, shared by some of the most prominent physicists of our time. Perhaps he may have been more right than we realise. We can't see ideas until certain conditions are met.

As we have discussed, a thought is a macro structure, but dependent on the presence of its constituents. This may be a little like appreciating JS Bach's Cello Suite No. 1–Prelude. Bach had to write it before you heard it, and it had to be a well written piece of music. The cello would have had to have been invented and become a popular instrument. You would have had to be in the right place at the right time to hear it played. You also need to be able to hear the frequencies of the musical sounds and the higher-order capacity to perceive its symmetry etc. (unlike baboons, who do not demonstrate an appreciation of classical music). We need to have the capacity of being able to write and read music, for Bach to have recorded it and in order for musicians today to be able to play the piece.

All these factors have to come together in the right way for one to hear the Bach piece and appreciate it. If just one of these factors is missing, appreciation is not possible. In the same way, if we do not appreciate or understand the components of what makes subspace travel possible, the idea of how they all fit together will not find fertile ground and be forever out of our reach.

It is therefore important not to allow the politics of knowledge to prevent us from collecting the information that must be combined in order to find mega-phenomena. If we treated the relationship between our knowledge disciplines differently, the answer may suddenly present itself, perhaps in the most unlikely of places. And we will think to ourselves, 'why didn't we see it before?—it seems so obvious'.[7]

In fact, the appearance of contradictory propositions or information is probably the best clue that a mega-phenomenon is being detected and waiting to be revealed.

There are tools already here to help us. TRIZ (in English, 'the theory of inventive problem solving', or TIPS) is a world-renowned problem-solving and innovation tool derived from the study of patterns of invention in the global patent literature. It was developed by the Soviet inventor and science-fiction author Genrich Altshuller. TRIZ embraces paradoxes. It identifies them as inherently solvable and builds our knowledge tools to help us think bigger than the 'problem'.

CONSCIOUSNESS, THE PRIME MOVER

Another possible scenario is that our evolving reality is facilitated by our consciousness. That is, instead of consciousness gradually releasing more and more of the rope of knowledge to humanity, our reality is sustained and created through consciousness. This is not a fringe view but one shared by prominent scholars across many different fields, from physicists like David Bohm and Max Planck, neurologists such as

Karl Pribram, psychiatrists like Stanislav Grof¹, biologists like Rupert Sheldrake, and engineers like Robert Jahn, Dean Radin and former astronaut Edgar Mitchell, the sixth person to walk on the moon. Lest we forget, it is also a view held by spiritual and religious leaders and arguably most consistent with the way the majority of the world—those from non-Western cultures—see existence.

However, there is a slight but important variation to this view being offered.

In the above example, when we think 'why didn't we see the answer before?—it seems so obvious', perhaps the answer is 'it wasn't there before; reality had to change in order to create it', which may include changes to the history of the universe as much as the future. For as Hawking suggests, '... we create history by our observation, rather than history creating us'.[8]

It may explain why science appears to follow our imaginations. Not long after George Lucas provides us with the image of a planet with two suns (in his first instalment of *Star Wars*), NASA scientists discover a planet with two suns. Science-fiction writers present to us the possibility of subspace and warp drives to harness it, and now our greatest minds lay down the theoretical foundations of space and time warps and NASA releases information that it is seriously investigating and researching warp-drive technology. HG Wells writes a book about a time machine, which is a hit around the world. Ten years later Einstein releases his theory of relativity, introducing the idea that time is not an absolute, but is malleable. There are numerous websites listing inventions inspired by *Star Trek*.

Coincidence? Perhaps. But it would not be the first time humans had observed life imitating art. Is there more to it? Is this another pattern we should investigate?

---

1 Stanislav Grof was chief of psychiatric research at the Maryland Psychiatric Research Centre and an assistant professor of psychiatry at the John Hopkins University School of Medicine.

It is possible that our progress is consciousness moving not only from the past to the future, but through different versions of the past and future? On day one, for example, we have X past and future. On day two we have a Y past and future. They both existed in space-time; however, consciousness created different ones. Is it possible that there is a mental arrow of causation as well as a chronological one (the one progressing 'forward' in space-time from the Big Bang to the present and into our future)?

In fact, many of the proponents of the holographic universe suggest that once one gets access to what physicist David Bohm terms the *implicate* order (or subspace) one has access to the past, present and future, as it all resides there in an integrated matrix. As the placebo effect, for example, provides evidence of our ability to access and manipulate the implicate order/subspace, there is reason to suspect that we may be carrying out similar manipulations of time and space also, as a hidden part of our everyday lives.

Could it be that Penrose and Hameroff are right about our consciousness representing our link to quantum reality, providing us with a cosmic presence that permeates the universe? Together with the brain, it could be the real quantum computer.

The 1960 movie THE TIME MACHINE, *starring Rod Taylor. Based on HG Wells's novel. Wells was born in the London suburb of Bromley in 1866, and began his literary career in 1895 with the publication of his first novel,* THE TIME MACHINE. *Ten years later Einstein releases his theory of relativity, introducing the idea that time is not an absolute, but is malleable.*

STAR TREK: THE NEXT GENERATION: *Captain Jean-Luc Picard in his quarters on the USS Enterprise (NCC-1701-D). He and his ship travelled and communicated through subspace, sometimes by the actions of others. Star Trek is the inspiration behind many of today's high-tech devices from mobile phones to MRI machines. Source: Derek Springer and Patrick Caughey collage. Effect added.*

Could group (or shared) consciousness select which reality to make a common reality? Does the popularity of *Star Trek* and *Star Wars*, for example, point to a reality that our shared consciousness might bring into being? Could our ancestors have been dealing with a *different* world, and a *different* reality, without the options we have today? Will tomorrow's scientific breakthroughs in warp technology or health mean that our history also needs to be changed in order to bring them into being (that is, history is changed to produce the right conditions to give us the future we want to create)?

These questions might sound outlandish, but there is no requirement for us to find reality mundane, conventional or predictable. Our future understanding may not be merely an extension of how we see things today. In time, evidence may show us a reality that is very different to our expectations. Besides, how much more outlandish is it, compared with the bizarre phenomena physicists are now telling us are indeed real. The real question which might confront us in the future is not *how* we can make things real with thoughts, but *why* we can't do this all the time.

A NEW PARADIGM

'Tell me,' the philosopher Ludwig Wittgenstein once asked a friend, 'why do people always say it was natural for a man to assume that the Sun went around the Earth, rather than that the Earth was rotating?' His friend replied, 'Well, obviously because it just looked as though the Sun was rotating?' Wittgenstein responded, 'Well, what would it have looked like if it had looked as if the Earth was rotating?'[9] As Wittgenstein suggests, perhaps we need to keep asking ourselves, what would our existence look like if the above suggestions about our reality are correct?

Some are arguing that we are long overdue for a scientific revolution; very little that has been revolutionary has happened in science since the 1970s.[10] People in the street, for example, can identify more Nobel laureates of the nineteenth and early twentieth centuries than they can laureates awarded since, at least in the sciences.

Academics like Gerald Pollack, Professor of Bioengineering at the University of Washington, an international leader in the field of water research, says this is because of the scorn and disdain of established science towards new ideas and/or challenging ones, mainly as a result of bureaucratic institutionalisation (people high in the establishment reward people who support their own beliefs), minutiae (people trying to become specialists in increasingly narrow areas) and vested interests (corruption).[11]

Pollack has teamed up with other notable scientists, such as NASA and Lockheed Martin Corporation aeronautics engineer James Ryder, to establish the Institute for Venture Science, an organisation set up to invest in 'promising ideas that challenge tired, worn-out paradigms'.[12] The institute is set up to back promising breakthrough ideas pursuing unexpected observations which would normally be passed over by established bureaucratic funding bodies wary of departing from the status quo.

The tree of specialisation makes it difficult for revolutions in scientific thinking to occur, the types instigated by Copernicus, Newton, Einstein, Wallace and Darwin, and Planck. Revolutions require the collaborations of different areas, difficult to achieve under our current, complex tree-like system of specialisation, with each branch and subbranch becoming more insular in its language and techniques.

The prevailing orthodoxy (in this case, the materialistic paradigm of nineteenth century naturalism) is therefore difficult to challenge. The suggestion is that any departures are hosed down in their early days, well before they are able to make links across branches. Einstein's achievements, for example, were remarkable as he was able to progress Special Relativity (which was much more than merely small alterations to the works of the theory founders at the time, such as Newton, Faraday and Maxwell) on his own without bouncing his early ideas off the scientific establishment. Indeed, if he had been a professional scientist, embedded in scientific institutions, rather than a clerk at a patent office, it is possible that his initial departures from the mainstream would have been quashed well before they were able to find the linkages they needed to develop into what they did.

Philosophers have been calling for the democratising of science for some time. Paul Feyerabend, Professor of Philosophy at the University of California, was known for his works *Against Method* and *Science in a Free Society*. He rejected the existence of universal methodologies (upon which much of contemporary science is based), providing examples of how adhering to them would have prevented many of history's great scientific breakthroughs, such has the Copernican revolution (which changed the paradigm of Earth being the centre of the universe).

He challenged the *consistency criterion*, the notion that new theories have to be consistent with old ones (which arguably would have prevented Special Relativity). He also challenged the *falsification* doctrine, where theories need to be consistent with all known facts, again arguing that many advances, such as those in quantum physics, would not have occurred if these rules were enforced. He called for the democratisation of science,[13] to free it from its constraints and allow greater non-professional participation. As we can see, Sheldrake and others are carrying on this tradition.

These voices have inspired 'citizen science' movements (which promote non-professional involvement in science) around the world, including in Australia, where on 6 May 2014 an inaugural workshop was held to support the development of the Citizen Science Network Australia,[li] following in the footsteps of established networks in Europe and the United States. SetiSETI@home, Galaxy Zoo, Wikipedia[lii] and

---

li *http://csna.gaiaresources.com.au/wordpress/?page_id=25*
lii *http://en.wikipedia.org/wiki/Citizen_science#cite_note-20*

even Internet search functions such as Google, are examples of citizen science ideas. Freeing science from the confines of institutionalism and enabling the broader community to engage with it may be what is needed to change our entrenched outdated paradigms. Think of the potential of having millions of minds working on a problem, instead of hundreds or even thousands (within the professions), examining it from very different perspectives and aptitudes!

New research by Sascha Friesike et al.[14] is illuminating the emergence of 'open science', based on '... the idea that scientific knowledge of all kinds should be openly shared as early as is practical in the discovery process'.[15] It is considered a paradigm shift in the treatment of research, away from the traditional practice of keeping research findings 'close to the chest', protected behind closed doors.

It is fostering a greater level of collaboration and innovation than the traditional scientific process and has been particularly successful in building partnerships between academia and industry. Academic institutions, for example, open up science by employing open-access journals and sharing research data. Large firms like Siemens, IBM or Tesla are also making their research publicly available. 'Instead of patenting knowledge, they publish large parts of their research in order to participate in the scientific community. In doing so they mark findings as state-of-the-art and thus prevent others from patenting them.'[16]

There are other positive signs that paradigms are beginning to change. Stephan Schwartz,[17] editor of *Schwartzreport*, an international online publication which provides information on trends in research around the world, thinks that a revolution is around the corner. Although sufficient evidence already exists in many different areas of science (he himself has first-hand involvement in researching the evidence for remote viewing), Schwartz thinks the research being produced now around near-death experience (NDE) will probably push our scientific paradigm over to a new evolutionary stage.

Schwartz says that current advances in resuscitation medicine, bringing patients back to life up to **four hours** after they have been defined as clinically dead, are providing opportunities to conduct prospective studies, that is, predictive research on NDEs.[18] The results are not only profound and significant, they are understandable to the person in the street, and for this reason they are less likely to be ignored in the way other anomalies in science are.

For Schwartz, the existence of a timeless realm able to provide us with insights confirms what creative geniuses (people who have become famous in history for their innovative and creative skills) such as Einstein, Michelangelo, Plato, Bach, and John Lennon have described, that is, an altered state of consciousness that connected them to another, timeless place where they *hear the music*.[19]

Schwartz sees our ability to access this dimension as hope for humanity as it struggles with contemporary challenges, such as the rising of sea levels, the destruction of the bees, the growing immunity of bacteria to antibiotics, and shortages of water. Here again we observe the value in accepting the validity of other states of consciousness in providing us information about reality.

Schwartz agrees that part of the problem is that 'science' is not a thing, but a series of 'silos' where disciplines develop 'their own languages', where 'physicians don't cite physicists' and 'physicists don't cite biologists'.[20]

Fragmentation is not unique to science and reflects broader social trends. Physicist David Bohm argues that our current way of fragmenting the world into parts not only doesn't work, but may lead to our extinction[21]. Montague Ullman, the founder of the dream laboratory at the Maimonides Medical Centre in Brooklyn, and a professor emeritus of clinical psychiatry at the Albert Einstein College of Medicine, New York, was responsible for ground-breaking research into the ability of dream states to show ESP ability and be used as therapeutic tools. His research pointed to the interconnectedness of human beings, supporting Carl Jung's concept of a 'collective unconscious', and says that 'unless we learn how to overcome all the ways we've fragmented the human race, nationally, religiously, economically ... we are going to continue to find ourselves in a position where we can accidentally destroy the whole picture'.[22]

Perhaps part of the answer is to build bridges between different disciplines. This will include developing modern science around spiritual intuitions, as we have done with our mechanical intuitions.

Supporting and continuing the meetings between Western science and the Dalai Lama is one example of good practice. The Clergy Letter Project, founded by Michael Zimmerman, is another endeavour, an international organisation of religious leaders and scientists created to demonstrate that religion and science need not be in conflict. The Clergy Letter Project sponsors the annual Evolution Weekend, an opportunity for congregations of all faiths to discuss the compatibility of religion and science.

In recent times, American Buddhist religious leaders have crafted a Buddhist Clergy Letter to parallel The Clergy Letters signed by more than 2,800 Christian clerics, 450 rabbis and more than 250 Unitarian Universalist ministers throughout the United States. Like the other letters, it makes it clear that deeply held religious belief can be fully compatible with the latest advances in science.

In fact, the preamble to the Buddhist Clergy Letter consists of a quotation from the Dalai Lama,

'If scientific analysis were conclusively to demonstrate certain claims in Buddhism to be false, then we must accept the findings of science and abandon those claims or adopt them as metaphor.'

Zimmerman says that from this perspective, the teachings of Buddhism, like the teachings of science, are provisional and must be considered open to change as new knowledge accumulates. 'The concept of metaphor is a critical one since so many religious leaders, like so many good writers, recognise the power metaphor has to drive home complex points. To read religious texts literally, to ignore the obvious metaphors present in each and every one, is equivalent to stripping the power from those texts. Why any religious person would want to do that is completely beyond my understanding.'[23]

And then there is *The Christian Science Monitor*, which prides itself on being an independent international news agency, with seven Pulitzer Prizes and more than a dozen international press club awards to show for it.

The Vatican Observatory and the University of Arizona are other good examples. The Vatican Observatory is one of the oldest astronomical research institutions in the world. Its research centre, the Vatican Observatory Research Group, is hosted by Steward Observatory at the University of Arizona, Tucson. The observatory conducts ongoing research and fosters international collaboration in astronomy to exploring the universe, as part of the Catholic Church's commitment to dialogue with the sciences.[24]

While I respect the perspective of the US National Academy of Sciences, suggesting that there are legitimately two ways of knowing, I'm not sure I agree that at no time shall their paths meet. Perhaps the time has come to start dismantling the apartheid.

The separation may have been useful at one point and perhaps served a purpose, like those other walls, such as the Great Wall of China, the Berlin Wall and the Israeli–West Bank barrier, the concrete wall currently separating the West Bank from Jewish settlements. But we intuitively know that these walls are not the answer and are built at least half with the hope that one day they will no longer be required; the day we have a more sophisticated understanding of how people can coexist. The same can be said of knowledge-gathering.

View from the West Berlin side of the Berlin Wall. The wall's 'death strip', on its eastern side, follows the curve of the Luisenstadt Canal (filled in 1932). This image was taken in 1986 by Thierry Noir at Bethaniendamm in Berlin-Kreuzberg.

Religion needs to do its part. It will need to better educate on the 'terrestrial' origins of religious writings. As Einstein, Hawking and Sheldrake remind science to avoid holding on too tightly to 'first principles' or the so called 'laws of science', religious leaders should also do their part to remind the faithful of how scriptures and the like were written by ordinary people trying to interpret what they saw and felt, largely through the eyes of their specific culture. As Zimmerman points out, metaphors are a useful tool for communicating complex points and need to be understood in this context. They should not be allowed to overshadow the points they were supposed to illustrate. Jesus, for example, made a point of reminding others, particularly Jews, of the dangers of holding on to laws (or 'first principles') too tightly. Indeed the New Testament was designed to free people from such laws.

We need also to be aware of how institutionalised religion can be a barrier to the pursuit of subspace. Buehler, for example, considers that religion can be an impediment to discovering the real subspace. For Christianity and Islam in particular, 'where religion has been heavily intertwined with politics, people thinking for themselves—bypassing the brokers of religion—is dangerous'.[25] Alternative paths to other realms of existence can challenge the authority of organised religions. In fact, as has been seen, efforts to find an independent interdimensional route, such as those of mediums, have been labelled by certain sectors of religious institutions as 'evil', in and of itself.

Creative geniuses on the spiritual side—such as Buddha, Jesus, Moses and Muhammad—would likely agree that the development of spirituality and religion should not, for some reason, stop with their teachings. In fact they picked up and developed ideas laid down by those before them, while also revolutionising them, in a similar way to Newton, Copernicus, Galileo, Wallace—and The Beatles, for that matter.

This does not diminish the importance of the older scripts. As discussed above, layering triangulation, for example, shows the value of perspectives gained at different points in time and that these perspectives represent valuable, unique and important information that must be protected.

Other interdisciplinary developments are also providing hope, such as current research into quantum biology, as discussed above. Here we are seeing 'biologists quoting physicists', as Schwarz might put it. A glimpse of what the future might hold.

In this new place, physics professors will be citing physicians, spiritualists will be citing biologists, and biologists spiritualists. The pictures of creative geniuses will be up on the same wall together: Jesus beside Darwin, Einstein beside Buddha, Planck beside Bach. As we have seen throughout this book, the really big names in discovery were those who were able to surf the connections between knowledge disciplines rather than deny them.

As we revolutionise our knowledge fields for the challenges ahead, we may find the limits of rationalism. We may also find that our everyday mental world, the landscape of thoughts we trek through daily, is an important connection to subspace. This inner world may currently be our best glimpse of what subspace could resemble. For some it might be frightening at first, but managing this fear may lead us to interesting places, not to mention possible answers to important questions.

The contradiction between the mechanical bottom-up view of reality and the view that supports the existence of a top-down force exhibited by the agency of our consciousness, might well be a tell-tale sign of a *mega-phenomena* greater and more profound than even subspace.

Whatever the case, one thing can be certain; the holy grail in science, the Theory of Everything (TOE), is incomplete without due consideration of the role of consciousness in explaining reality. Our conscious, subjective experience appears to be a component of everything, having as much to do with the creation of the material universe as with the non-material.

Sir John Charlton Polkinghorne was professor of mathematical physics at the University of Cambridge from 1968 to 1979, when he resigned his chair to study for the priesthood, becoming an ordained Anglican priest in 1982. He is the author of five books on physics, and 26 on the relationship between science and religion. His deep understanding of both religious and scientific thought allows him not only to understand but to explain the connections like few others can. His writings show how thinking more deeply can reconcile what can appear at a superficial level to be contradictory propositions. He and others like him (and there are a few) are probably some of our best guides when it comes to navigating the mega-phenomena of subspace.

Our safari into physics, biology and religion to *find the elephant* has shown the role of consciousness, from collapsing wave forms into matter (quantum physics) to connecting us to a greater timeless and spaceless plane of our existence (psi), to changing the chemistry of the body (placebo), to explaining life (evolution).

You don't have to understand complex quantum experiments to get this. Just consider your own memory. Who or what is it that decides what memories to recall? Who or what is examining them? That is you, the most profound mystery in the universe. Without accepting the role of our consciousness, subspace—and perhaps many more phenomena—will likely remain beyond our intellectual reach. The fifth dimension may well be the mental realm.

Despite the evidence gathering against it, including the developments in NDE, it is likely that the materialistic paradigm of nineteenth century naturalism will continue to be championed by significant portions of professional science for some time yet. However, as the evidence mounts and eventually draws in the new generation of thinkers and scientific explorers, some will decide that materialist commentators no longer have anything valuable to contribute and should be seated with flat-earth advocates at the 'to be ignored but historically interesting' table.

But, of course, that would be wrong, as the whole point of this book was to show that everyone's perspective not only deserves to be heard but *needs* to be heard when we are on the hunt for complex and broad phenomena (mega-phenomena) that can save our collective skins. Challenging the materialistic paradigm was designed to allow room for other perspectives in the scientific process, not cleanse science of what appears to be outdated views. Indeed, determinism has its place, as it enables us to have confidence in our space-time predictions. Moreover, while we have established that determinism can't explain our existence or our ability to interfere with the deterministic unravelling of space-time, its relevance to subspace itself remains unclear.

It is helpful now to look at models or frameworks that could help us bring together the different aspects that make up reality, both the material and the non-material, in a practical way. We have mentioned

TRIZ but following are two other popular theories that might help provide such frameworks or models.

NEW MODELS FOR OUR NEW PARADIGMS

Is it all just a hologram?

As mentioned in the physics section, the mathematics of black holes suggests that the universe may be a giant hologram. The idea that the universe could be a hologram was most notably launched by Michael Talbot's classic in the field of science and spirituality, *The Holographic Universe*.

A hologram is a three-dimensional image made up of light-wave interference patters, much like the message RD D2 delivers to Obi-wan Kenobi from Princess Leia (in Obi-wan's cave) during the first instalment of George Lucas's *Star Wars* movies. The hologram was invented by a brilliant Hungarian-born physicist and inventor, Dennis Gabor, in the 1950s. In 1971 Gabor won the Nobel Prize in Physics for his development of the holographic method.

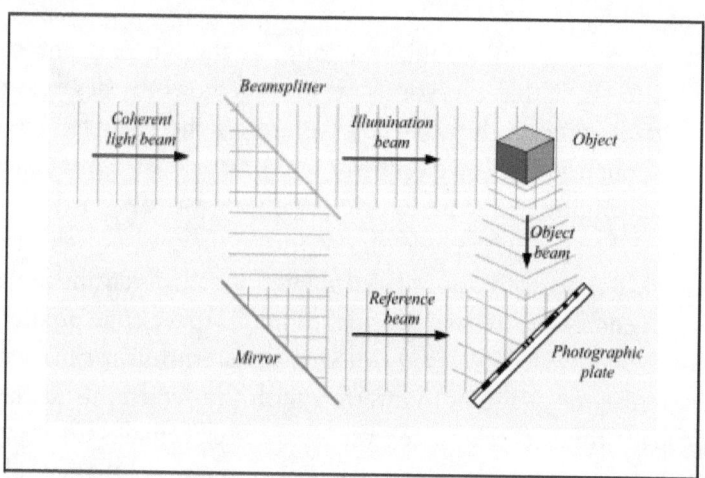

Simply, a hologram is created when a laser beam (coherent light beam) is split into two. One half goes via mirrors directly to a photographic plate (reference beam) and the other (illumination beam) to an object one wants to make a hologram out of. The laser light is reflected off this object onto the same spot on the photographic plate.

The two beams intersect and interfere with each other. The interference pattern (caused by the two different wave patterns coming together) is imprinted/recorded on the photographic plate.

Shining a laser onto the photographic plate (reconstruction beam) will create the virtual three-dimensional image in space for the observer.

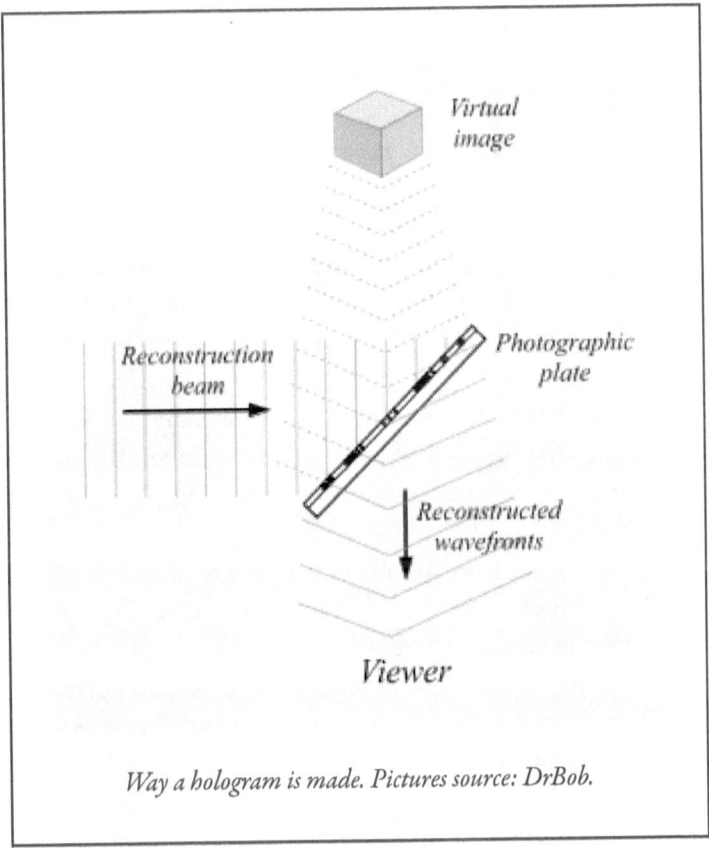

Way a hologram is made. Pictures source: DrBob.

One of the key differences between a photograph and a hologram (besides its obvious ability to produce actual 3D images) is that if one cuts a photograph in half, one only gets half the picture on each side. However, in a hologram every part of the 'picture' (in the photographic plate pattern) contains all of the information of the picture. Every portion of the holographic film contains all of the information of the whole. Shining a laser through only a portion of it may provide a fainter image, but an accurate one of the whole image nevertheless. Also, because of its multi-dimensions, a hologram has huge potential for information storage per square millimetre.

Talbot built on the research of University of London physicist David Bohm (a protégé of Einstein and one of the worlds most respected quantum physicists) and Stanford neuroscientist Karl Pribram, author of the classic neuropsychological textbook, *Languages of the Brain*. Both interpreted twentieth century anomalies in physics and in neuroscience as clues that the universe (all matter and space, including us) is really a hologram projected from another, more primal dimension of reality.

Pribram began considering anomalies in current thinking about brain mechanics while working with neuropsychologist Karl Lashley. For over thirty years Lashley researched the mechanics of memory. He was trying to find out where memory was stored in the brain but was confounded at every turn by tests that invariably showed that no matter which parts of the test rats' brains were removed, the rats still remembered their trained tasks. In his own research, Pribram observed the same phenomena in persons who had parts of their brains removed due to accidents and surgery without suffering significant changes to their memory.

Similar effects were also found in relation to vision. For example, after removing ninety per cent of a rat's visual cortex (the part of the brain that receives and interprets what the eye sees), Lashley found the rat would still perform trained tasks requiring complex visual skills. Pribram's own research also found that cats could have ninety-eight per cent of their optic nerves severed but still be able to perform complex visual tasks.[26] This provided support for the view that certain functions in the brain are not location-specific, that information may be stored in a manner that is not consistent with the mechanical brain models.

Pribram also noticed that our experience of the world is not (as Sheldrake also suggests) in our skulls, or even in our eyes (where we perceive it), but 'out there'. When we have a sore finger, we feel it in our finger, not our brain. In the 1960s Nobel Prize winning physiologist Georg von Békésy placed vibrators on the knees of blindfolded test subjects and varied the rates of the vibrations. He discovered that he could make his subjects feel sensations on different knees, even between their knees, where they had no limbs or sensors.[27] It showed that our brains could create our spatial experiences through the analysis of wave forms.

These and other puzzling properties of the brain—including the remarkable amount of information it can store, its 'plasticity' (as discussed in a previous section, which undercuts classical models of brain function[liii]), and even the difficulty researchers have in identifying where the mind begins and ends in the body—led Pribram to emerging holographic research. Perhaps the brain stores information like a hologram, as records of wave-interference patterns. It could be an evolutionary mitigation of the catastrophic risk of losing critical portions of information as a result of neurological damage to any one specific section of the brain. What if a micro-neurological accident in a small portion of his brain meant that Mozart just lost his knowledge of the note A#? It would be devastating. What if an Olympic runner suddenly forgot how to stop running? What if a father had a small neurological accident that meant he remembered everyone in his family except one of his daughters? This kind of thing doesn't happen much, if at all.

Pribram's views were supported by contemporary research in physics, particularly by individuals like David Bohm, one of the founding academics who suggested 'hard' reality did not exist, that underlying the material universe we are familiar with is a complex array of wave-like interference patterns which only become solid to our sensors when we observe them. As the world was uncovering the non-local

---

liii Talbot gives an example of an exercise where you must try to trace your first name in the air with your elbow. The relative ease with which people can do this, even for the first time, contradicts the classical view that various areas of the brain are 'hard-wired', only able to perform tasks *after* repetitive learning has caused the neuro-connections to become established (Talbot, 1992, p. 24).

properties of the universe, such as the phenomena of entanglement and the malleability of time and space, Bohm, too, looked to the hologram as a more accurate model, but of the mechanics of the universe rather than just the brain, explaining how what appears to be separate on one level is really connected at a deeper one. Bohm proposed that the universe was a holographic projection from another dimension, an *implicate order* ('enfolded' state). Within this implicate state all things are connected, including us, our consciousness, matter, space and time. Indeed, matter and consciousness are considered by Bohm as different points on the same continuum, one an extension of the other.

University of London physicist Professor David Bohm (a protégé of Einstein and one of the world's most respected quantum physicists) is one of the founding academics who suggested 'hard' reality did not exist, that underlying the material universe is an array of wave-interference patterns. Source Karol Langner.

For both Bohm and Pribram, anomalous findings in other emerging areas—such as research into the placebo effect, near-death experience, and psi ability (including ESP, telekinesis and remote viewing)—add weight to the holographic model. They showed that our consciousness had an ability, under the right circumstances, to connect with the implicate order and affect it, resulting in changes to our everyday world which Bohm refers to as the *explicate order* (or 'unfolded' order), in ways that would appear miraculous to many of us.

The possibility that the universe we experience is a hologram is a view that is now taken quite seriously in science. For example, physicist William Tiller,[liv] former head of the Department of Material Science at Stanford University, agrees with the hologram view, equating reality to the holodeck on the television series *Star Trek: The Next Generation*.[28] Though for our investigation the key point is that there is clear and significant support for the existence of another dimension in addition to our everyday ones, a dimension integral to our experience of reality, where time and space obey different laws, but one we know far too little about. While it is a dimension that is largely ignored by Western science more and more scholars are now prepared to cry out that indeed, 'the emperor has no clothes.' Furthermore, it is not beyond our control or our ability to test, and is awaiting our more serious efforts to harness its potential. The holographic model provides us with one possible framework for us to work with.

A new integrated model—integral theory.

To capture meta-phenomena perhaps we need *mega-paradigms* able to bring together the different disciplines in a practical way to meet today's complex and often global challenges.

One model has surfaced that brings the different human disciplines of knowledge together, including the natural and social sciences as well as the arts and humanities. It is based on American philosopher Ken Wilber's *integral theory* and is now being used in over thirty-five distinct academic and professional fields such as art, international development, health care, organisational management, ecology, economics, psychotherapy and law.[29]

Integral theory is sometimes referred to as the theory of everything, from the title of Wilber's 2000 book, *A Theory of Everything: An Integral Vision for Business, Politics, Science, and Spirituality.*

---

liv Founder of the Institute of Psychoenergetic Science
http://www.tillerinstitute.com/index.html

I Interior-Individual Intentional Subjective **Upper Left**	**IT** Exterior-Individual Behavioral Objective **Upper Right**
WE Interior-Collective Cultural Intersubjective **Lower Left**	**ITS** Exterior-Collective Social Interobjective **Lower Right**

Ken Wilber's four quadrants express the four areas of reality which need to be considered in order to understand what is happening in any situation.

Integral theory has become one of the dominant approaches in meta-theory and has proven a powerful framework able to be applied to virtually any context, from one-to-one psychotherapy to the World Bank's Leadership for Results program.

According to this theory, there are at least four irreducible perspectives on, or dimensions of, reality that must be considered when we attempt to understand any issue or phenomenon. It proposes an *inside* and *outside* perspective, as well as a *singular* and *plural* perspective.

Wilber at times considers the Upper Right (IT–UR) and Lower Right (ITS–LR) a singular objective sphere, in contrast to the subjective Upper Left (I–UL) and intersubjective Lower Left (WE–LL) spheres. He equates theses spheres with other similar approaches to reality in history, such as those proposed by Plato and others,[lv] that is, the spheres of the *True* (objective, IT and ITS), *Beautiful* (subjective I), and *Good* (intersubjective WE).[30]

Unlike other models, this acknowledges the complexity of reality and does not try to reduce its analysis to merely one of the quadrants. It allows an appreciation of a situation through these different lenses. It examines how one experiences something (the I), the behaviour exhibited (the IT), the cultural attributes related to it (the WE) and the social/systemic aspects (the ITS).

[lv] Immanuel Kant's spheres of *Judgement, Practical Judgement* and *Reason* and Jürgen Habermas's *Truth, Rightness* and *Truthfulness*.

Let's take an example. Consider being confronted by a snake while on a bushwalk. To understand the dimensions at play in this situation, one can consider how you might feel when confronted by the animal. You might experience a sense of fear and you may decide to stop and be still (the I).

I	IT
Interior-Individual	Exterior-Individual
Intentional	Behavioral
Subjective	Objective
Upper Left	**Upper Right**
WE	**ITS**
Interior-Collective	Exterior-Collective
Cultural	Social
Intersubjective	Interobjective
Lower Left	**Lower Right**

Another dimension to consider is the cultural aspect. What does our culture say about snakes and how we should relate to them? This might, for example, suggest that snakes are commonly considered dangerous or even evil. There could be cultural rules about how we should treat them. There may be suggestions that being confronted by a snake may be an omen of some sort (the WE).

Then there will always be the specific objective details of the situation, such as the physiology of the snake, its type, its chemical make-up, the potency of its venom, its actual behaviour and its distance away from you etc. There are also the chemical reactions occurring in your brain when confronted by the snake. These are the objective aspects (the IT).

There is also the natural environment, the snake's place in it and the local ecology, as well as our own interaction with this ecology, or the web of life. Also of relevance are government laws about the treatment of flora and fauna, and market forces which may make catching such a creature a money-making venture. These factors relate to the objective systemic (plural) aspects of the situation—that is, the ITS.

These systems are interconnected and integral to any situation. Wilber suggests they are born together.

We can use the quadrant model in two ways: either to map the *dimensions* (Wilber refers to this as the *quadratic* approach) of a situation, or to map our *perspectives* on, or ways of seeing, a situation (*quadrivia*).

I Interior-Individual Intentional Subjective **Upper Left**	**IT** Exterior-Individual Behavioral Objective **Upper Right**
WE Interior-Collective Cultural Intersubjective **Lower Left**	**ITS** Exterior-Collective Social Interobjective **Lower Right**

Wilber also suggests that the left side of the quadrant (I and WE) can range in levels of *depth*, while the right side (IT and ITS) ranges in levels of *complexity*. The levels in each quadrant enable us to see how that quadrant is expressed in its different manifestations.

Within each quadrant Wilber shows correlating *lines* of development that correlate to lines in the other quadrant. These lines might, for example, indicate the different kinds of intelligences each person has (cognitive, emotional, moral) (I), or lines of development from the family, to the group to the culture to the societal (WE).

There are also *states*, temporary expressions or occurrences of reality that are incompatible. For example, one can't be happy and sad at the same time.

Alternatively there are *types*, consistent styles that arise irrespective of developmental levels. These might include gender (I), blood types (IT), religious systems (WE) or political regimes (such as democracy or communism) (ITS).

There are *zones* as well, which are different ways of *knowing*. There are eight zones (two in each quadrant), including first-person realities, (phenomenology) (I), second-person realities (ethnomethodology) (WE), and third-person empirical realities (empiricism) (IT).

Integral theory uses colours to represent each distinct level (for example, red, amber, orange, green, teal, turquoise), which also represent the general movement of a widening identity, from the 'me' to 'my group', to 'my country' to 'all of us', to the planet right through to the *Kosmoscentric* view.[31]

I	IT
Interior-Individual	Exterior-Individual
Intentional	Behavioral
Subjective	Objective
Upper Left	**Upper Right**
WE	**ITS**
Interior-Collective	Exterior-Collective
Cultural	Social
Intersubjective	Interobjective
Lower Left	**Lower Right**

All these are mapped in Wilber's AQAL Integral Map.[lvi]

Wilber adopts the *Kosmos* perspective on reality as defined by the Greeks, that is, 'the patterned Whole of all existence, including the physical, emotional, mental and spiritual realms. Ultimate reality was not merely the cosmos, or the physical dimension, but the Kosmos, or the physical and emotional and mental and spiritual dimensions altogether. Not just matter, lifeless and insentient, but the living Totality of matter, body, mind, soul, and spirit.'[32]

As his model acknowledges, the place of consciousness in explaining reality it is arguably superior to other 'theories of everything', such as M-theory, which dwell in Wilber's right-hand quadrants, that is, the objective singular and perhaps objective plural. It finds a place for the array of phenomena exhibited in this book which the descendants of the reductionist, nineteenth century school of science find so difficult to explain.

Wilber himself is another who feels that the world is on the precipice of a monumental paradigm shift. In *A Brief History of Integral with Ken Wilber*,[33] Wilber talks about how humankind has undergone five major technological sequential transformations:

---

lvi See *https://integrallife.com/integral-post/overview-integral-theory*

- foraging (hunt and search for food),

- horticultural (growing food),

- agrarian (cultivating the land),

- industrial (manufacturing), to

- informational (computer, digital, new media age).

These correlate with the progression of world views:

- archaic (instinctual drives and needs).

- warrier magic (pagan gods and traditions),

- traditional mythic (religious),

- modern rational (scientific), to

- postmodern pluralistic (multiculturalism).

In that order. Each structure transcends and includes the former.

In today's Western world, says Wilber, about thirty per cent of the population is traditional mythic (fundamentally religious), forty per cent is modern rational (scientific) and about twenty-five per cent is postmodern (multicultural). Individuals are born at the archaic and grow and develop through the higher stages. They may stop at a particular stage for a while.

Wilber says we are in the midst of another worldwide transformation, which is unprecedented. This is because all previous stages considered that they—and they alone—held the correct view. A zero-sum scenario operated; that is, one could only be right if another was wrong. Wilber sees our ever-present culture wars as the result of feuding between these major stages (mythic religious, rational science and post-rationalism/postmodernism).

The coming stage is regularly termed *integral*. In this stage, room is made for all of the stages. Each is true but partial. Each is more inclusive than the former. The integral stage is the first stage in all of human history that is all-inclusive. He sees this stage as potentially replacing conflict with harmony for the first time in human history.

Wilber suggests that currently about five per cent of the population is at the integral stage. History shows that when approximately ten per cent of the population reaches the leading edge of evolutionary values, the entire culture starts to adopt the new values. Wilber maintains that this can be seen in all the previous stages.

When integral values are adopted by ten per cent of the population, Wilber believes it will begin a landslide. The integral view will permeate through world culture and bring forward profound changes in humanity. He suggests this is coming 'just in time', as the worldwide problems of today's generation—such as global warming and terrorism—require an all-encompassing framework to be addressed. One nation or world view acting alone is insufficient to solve these challenges.

Integral theory is showing great promise, increasingly being adopted in different disciplines internationally. The Integral Research Centre, set up by Wilber in 1998, undertakes integration research in many fields. The peer-reviewed *Journal of Integral Theory and Practice* publishes academic articles and case studies. The John F. Kennedy University's Department of Integral Theory has an online master's degree in integral theory. Indeed, Integral Coaching Canada, an Ottawa-based company, has developed an entire school and methodology for professional coaching based on the AQAL model. There is also the international biennial Integral Theory Conference, the fourth of which is being held as I type these words at the Sonoma State University, California.

Like any innovation, this theory raises some questions. For example, in a post-rational world, is a detailed, even 'mathematical', model or map going to be effective? We are attempting to explain a reality from which mathematics and form have sprung.

Also, there is a suggestion that each evolutionary stage is superior to the one before; for example, the multicultural world view is superior to the scientific or religious world view. When someone moves from one to the other, it shows a sign of 'development'. This may come across as too judgemental for some.

There is one interesting experience of Wilber's that is helpful to remember, going forward. In a talk, Wilber suggested that he had the impression that when he was able to show value in uniting the different world views there would be some gratitude shown by proponents of each view. To his surprise, he experienced the reverse. Proponents of the different world views disliked his vision intensely.[34]

On reflection, Wilber realised it was because each view is founded on a belief that it, and it alone, is correct. While Wilber acknowledged the value of their perspectives, the fact that he also acknowledged the perspectives of all the views was a deal-breaker.

Those of us wishing to promote a more integrated view of reality will need to be prepared for negative reaction from proponents of other world views, still operating within a zero-sum framework. Particularly vitriolic are likely to be those who have made a living in promoting their perspective. The best of luck to you!

THE ROLE OF THE OUTSIDER

Triangulation has the capacity to explore the merits of housing different perspectives, views, insights, skills and experiences to a greater extent than described here. It brings together the 'educated' and 'uneducated', the 'bright' and the 'dull', 'conservatives' and 'liberals'. It provides solid evidence of the value of accommodating different views.

Triangulation, models like Wilber's and initiatives like the Institute for Venture Science have the potential to open up science not only for academics, but for non-professional enthusiasts. This is not a bad thing. After all, neither Darwin nor Einstein were professional scientists when they devised their seminal theories. How many Darwins and Einsteins has the world been denied by modern science's institutionalised aloofness?

Indeed, it could be that the so-called amateur's ability to see things from a perspective outside the view of our institutionalised silos has an important role to play in spotting phenomena that sprawl across the divides which professionals have created. It is a role for the 'outsider', and possibly an antidote to our over-specialised knowledge architecture. As Allan Snyder, of Sydney University, put it, people are blinded by their expertise.

As each of us might have a place in finding answers to our greatest questions, it brings with it the possibility that we all might be able to understand the answers. If our minds can ask the question, I strongly suspect our minds will also be able to comprehend the answer. For if we can identify a hole, this implies we should also be able to know when it is filled. We don't need to become particle physicists, cave-dwelling monks, molecular biologists or spiritual mediums to 'get it'.

Allen Carr was an Englishman responsible for helping, to date, what some estimate to be between ten million and twenty-five million people break their addiction to cigarettes. His method, which is inexpensive and uses no drugs or scare tactics, achieves results that put to shame public health campaigns like Quit.[35]

After fifteen years of smoking, I had tried everything to quit; cold turkey, hypnotherapy, nicotine gum, the lot. I had heard of a book that colleagues said worked like magic. I dismissed it, of course, and kept on trying conventional methods. If the book was any good, I would be hearing about it from health professionals or the government, right?

The problem was that after years of trying the conventional methods, I was still smoking. Nothing seemed to work for me. In desperation, I finally decided to track down this book I had heard so much about. It was Allen Carr's *The Only Way to Stop Smoking Permanently*.

It was one of the easiest methods I had tried and now, after nearly another fifteen years of not smoking, I can say with confidence that it was the most successful. Although clinics are offered, most people I know who used Carr's technique, like me, achieved success just by reading one of his books.

You would be forgiven for thinking Allen Carr was a hypnotherapist, psychologist, psychiatrist, medical practitioner, marketer, academic, researcher, drug and alcohol worker or some other health professional when he discovered his cure. He was actually none of these; he was a regular accountant with a hundred-cigarettes-a-day habit, the result of twenty-three years of addiction. At the age of 48 he was desperately trying to find a method to stop smoking that worked for him. On 15 July 1983, while struggling to read the biochemical section of a medical book, his life experience melded with what he was reading in a unique way: he had a profound insight into the phenomenon of nicotine addiction.[36]

Once he saw it, he stopped smoking. He wrote it down and published it. It turned conventional thinking about nicotine addiction upside down. Many agreed with Laurence Phelan of *The Independent*, that 'the active ingredient in Carr's cure is a paradigm shift; an upending of perceived wisdoms'.[37] Thankfully for me, Phelan, Sir Richard Branson, Pink, Sir Anthony Hopkins, Ellen DeGeneres, Lou Reed, Anjelica Huston, Ashton Kutcher and millions more like us, Carr did not let his unqualified status (in the eyes of the establishment) prevent him telling the world about what he had found.

His approach rests on his unique understanding of what happens to the body when nicotine leaves it, that is, it creates an empty, insecure feeling. The next cigarette brings back some of the confidence it took. In his own way, he had *found the elephant* amongst the plethora of competing remedies and approaches to cigarette addiction he had experienced, and had the fortitude to stand up for it. Through his unconventional but simple method, Carr has prevented so much human suffering and saved so many public health dollars, that I consider him an example of how one person can change the world.

And then there is Sam Tsemberis. You may not have heard of Sam but, according to health professionals and academics across the United States, this 'outsider' has stumbled across an approach that has all but solved one of society's most intractable social problems; that is, chronic homelessness.

His recent test run nearly eradicated chronic homelessness in Utah. In Phoenix, an earlier test case eliminated chronic homelessness among veterans. Then New Orleans housed every homeless veteran.[38]

Tsemberis says he was never trained in how to treat the homeless. 'I'm a psychologist,' he said in an interview with Terrence McCoy of *The Washington Post*.[39]

Born in Greece and raised in Montreal, Tsemberis took a job in New York City in the early nineties doing outreach for the mentally ill. This brought him into close contact with the homeless. Getting to know them, Tsemberis realised that the way some experts perceived homelessness was fundamentally flawed. He observed that survival in the streets was much more laborious and complicated than people understood, and required considerable skill.

Housing people as a reward for their attendance at counselling sessions and their conquering of addictions was 'putting the cart before the horse'. He saw a need for change, and assembled a small team who, like him, had no training in homelessness. It included a recovering heroin addict, a formerly homeless person, a psychologist, a poet and survivor of incest.

Give homes to the homeless and you will solve chronic homelessness, he said. The approach consists of prioritising the chronically homeless: those with mental or physical disabilities who are homeless for longer than a year and who require the most government expenditure for support (in hospitals, jails and shelters).

Then give them a home, no questions asked. Immediately afterwards, provide counselling. Give them final say in everything, from where they live to how they are counselled.

This may sound simple—and, for some, counterintuitive—but it appears to work. Research has shown that the program typically achieves around eighty-five per cent retention rates, compared with the next-best model, which achieves around sixty per cent. The Tsemberis model is not only more effective but is slashing millions of dollars off the government costs of servicing the needs of the homeless.

Tsemberis was able to connect the dots where the institutionalised response became too politicised and siloed to 'see the forest for the trees'. With his fresh eyes on the problem and a position outside of the establishment, he was able to see what so many of the world's experts either could not or would not. He is yet another example of the role 'outsiders' have to play as an antidote to the failings of our over-specialised and politicised knowledge industry. He was able to use his unique insights and fresh perspective to come up with an answer for a problem that many had given up on solving.

Of course, there are more examples of the outsider's contribution. And at some future time these examples may even include your story.

So now it is over to you. Your unique biology and life experiences give you your own perspective on the world. No one sees it quite the way you do. While this means that others can see some things better than you, it also means there are things you can see better than others, even so-called experts.

What mega-phenomena can you see that others can't? What parts of your experience and knowledge can you bring together with that of others to reveal new insights? It might be an answer to a social problem, or an environmental one. It might relate to a new challenge, such as global warming, or one that has dogged humanity for some time, like the Arab-Israel conflict or how a bicycle works. It could be an answer to a public health issue, as Allen Carr had discovered. It could be a local issue or one no one has recognised yet. There are plenty of intractable problems left in this world for you to choose from.

The character Uncle Ben said to Peter Parker in *Spider-Man* (the 2002 movie), 'with great power comes great responsibility'. This may also imply that with unique power comes unique responsibility. Perhaps we all have our elephants to find.

APPENDIX

DELAYED-CHOICE QUANTUM ERASER

This experiment is known as one confirming that equipment is not to blame for the observer effect and that, again, information is communicated back in history.

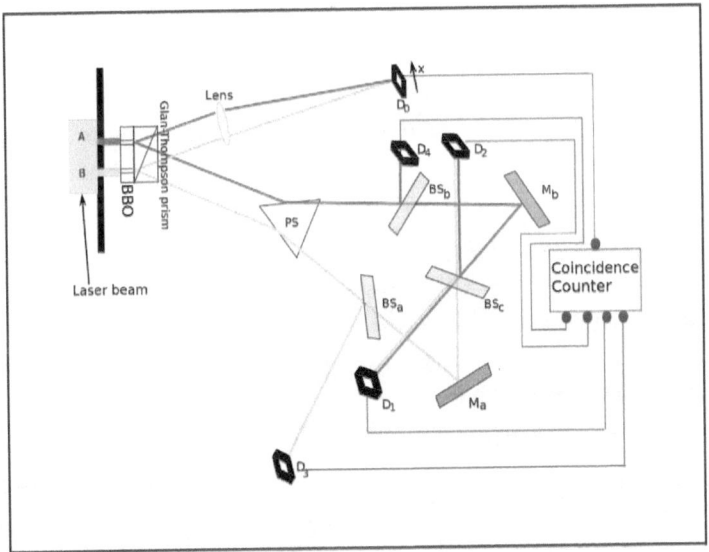

In this experiment[1] a photon is generated and passes through a double-slit apparatus (vertical black line in the upper left-hand corner of the diagram).

The photon goes through one (or both) of the two slits. Its paths are shown as red or light blue lines (red indicates slit A, light blue indicates slit B)(for those seeing a black and white diagram, dark gray indicates slit A, light gray indicates slit B).

After the slits a special crystal (beta barium borate crystal—labelled as BBO) converts the photon into two identical entangled (as they were once together) photons with half the frequency of the original photon. These photons are caused to diverge and follow two paths by the Glan-Thompson prism.

The photons that go north-east to the D_0 detector are referred to as the *signal* photons.

The other entangled photons, moving south-east, are referred to as the *idler* photons.

Later on the idler path, beam splitters (green/light grey blocks) randomly allow the photon to either pass through or be deflected. The gray/dark gray blocks in the diagram are mirrors.

Because of the way the beam splitters are arranged, the idler can be detected by detectors labelled D_1, D_2, D_3 and D_4.

If the photon is recorded at detector D_3, then it can only have come from slit B.

If it is recorded at detector D_4 it can only have come from slit A.

If the idler is detected at detector D_1 or D_2, it might have come from either slit (A or B).

Detectors of the idler provide information as to which slit the photon came from—that is, A or B (at detectors D_3 and D_4)—or that the information is unknown (D_1 and D_2).

At the later stage (at D_1 and D_2), the detectors are unable to provide information on which slit the photon came from, and so at this point the path information is termed to have been *erased*.

When the experimenters looked only at the signal photons whose entangled idlers were detected at D_1 or D_2, they found an interference pattern.

However, when they looked at the signal photons whose entangled idlers were detected at D_3 or similarly at D_4, they found no interference.

Here we see that the choice of whether to preserve or erase the path information of the idler is made *after* the position of the signal photon has already been measured by D_0. Thus the information at D_1 and D_2, as well as at D_3 and D_4 (at points in time after the signal photon has

APPENDIX

hit D_o, due to the shorter optical path) determines whether interference is seen at D_o or not.

Again this suggests a movement back in time, through a subspace-like dimension not restricted by time and space.

The explanation again in time line

It is worth getting your head around this experiment. In many ways it describes the profoundness of quantum physics and will likely change your view about the world we live in, particularly if you have been raised in the West.

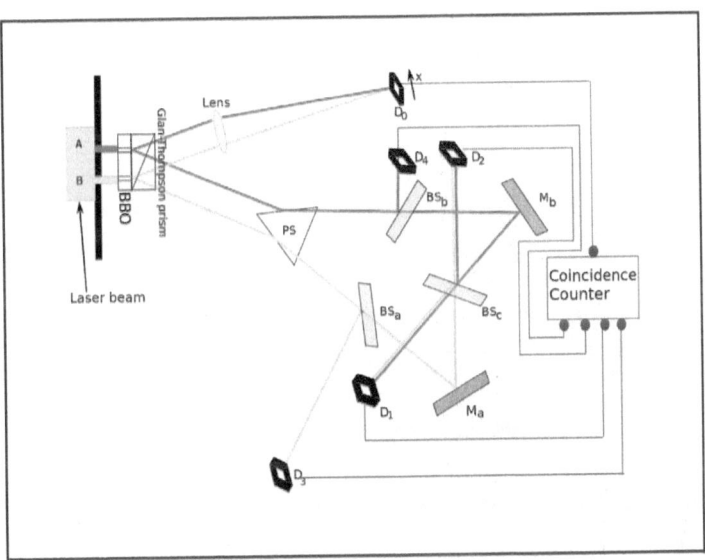

So I provide another way of describing what is happening, which I discovered when reviewing commentary by Ross Rhodes,[2] through a time-line explanation.

Time 1. The entangled pair leaves either region A or region B of the crystal. The signal photon heads off to detector D_o, and the idler photon heads off to the interferometer (PS).

Time 2. The signal photon is registered and scanned at detector D_o according to its position.

This information (the position of the signal photon upon 'impact' at D_o) is sent on its way to the Coincidence Counter.

Time 3. The idler photon reaches the first pair of beam splitters, BSa, BSb. There, a random choice on which direction the idler photon will go—either to detectors D_3, D_4, or to the quantum eraser BSc and on to detectors D_1, D_2.

Time 4a. If the idler photon is sent to detectors D_3, D_4, it is detected with which-path information intact. Then, and only then, do we know which-path information for its twin signal photon, which has already been detected, scanned, registered and recorded at D_0.

Time 4b. If the idler photon passes through to detectors D_1, D_2, it is detected with no which-path information (the which-path information having been *erased* at BSc).

Time 5. The Coincidence Counter correlates the arrival of a signal photon at detector D_0 with the arrival of its twin at D_1, D_2, D_3, or D_4. If the correlation is with an idler arriving at D_3 or D_4, then we know (after the fact) the which-path information of the signal photon that arrived earlier at D_0.

If the correlation is with an idler arriving at D_1 or D_2, then we have no which-path information for the signal photon that arrived earlier at D_0.

Time 6. Upon accessing the information gathered by the Coincidence Counter, we the observers are shocked to learn that the pattern shown by the positions registered at D_0 at Time 2 depends entirely on the information gathered later at Time 4 and available to us at the conclusion of the experiment.

The position of a photon at detector D_0 has been registered and scanned. Yet the actual position of the photon arriving at D_0 will be at one place if we later learn more information, and will be at another place if we do not.

Both the above experiments produce findings that, under a naturalistic or classical view of reality, are impossible. Nevertheless, the results have been consistent time and time again and are not seriously questioned.

BIBLIOGRAPHY

1. Adler, R. *The many faces of the multiverse. New Scientist*, 25 November 2011. http://www.newscientist.com/article/mg21228402.200-ultimate-guide-to-the-multiverse.html

2. M. Arndt, O. Nairz, J. Voss-Andreae, C. Keller, G. van der Zouw, and A. Zeilinger, *Wave-particle duality of C60 molecules, Nature*, London. http://physicsworld.com/cws/article/news/1999/oct/15/wave-particle-duality-seen-in-carbon-60-molecules

3. Barrett, J. *The God Issue: We are all born believers, New Scientist*, 21 March 2012. http://www.newscientist.com/articleimages/mg21328562.000/1-the-god-issue-we-are-all-born-believers.html

4. Botton, A. *Religion for Atheists, A non-believer's guide to the uses of religion*, Penguin Books, London, 2012.

5. Carey, B. *Antidepressant Studies Unpublished, The New York Times*, 17 January 2008 http://www.nytimes.com/2008/01/17/health/17depress.html?_r=1

6. Carey, B. *Paralysed, Moving a Robot with Their Minds, The New York Times*, 16 May 2012. http://www.nytimes.com/2012/05/17/science/bodies-inert-they-moved-a-robot-with-their-minds.html?_r=3

7. Carter, C. *Does Telepathy Conflict with Science? Many are starting to think not, The Epoch Times*, 2012. http://www.theepochtimes.com/n2/science/does-telepathy-conflict-with-science-211214.html

8. Cecile A. et al. (J.W. Janssens, Marta Gwinn, Linda A. Bradley, Ben A. Oostra, Cornelia M. van Duijn, and Muin J. Khoury *A) Critical Appraisal of the Scientific Basis of Commercial Genomic Profiles Used to Assess Health Risks and Personalise Health Interventions. The American Society of Human Genetics.* Published by Elsevier Ltd 2008. http://www.ncbi.nlm.nih.gov/pmc/articles/PMC2427295/?tool=pubmed

9. Cheetham, N. *Introducing Biological Energetics; how energy and information control the living world.* Oxford University Press, 2011.

10. Cheetham, J. *The Dawkins Dilemma, The Sun-Herald,* 7 March 2010. http://www.jockcheetham.com/OneWorld/Home_files/The%20Dawkins%20Dilemma%20by%20Jock.pdf

11. Clayton, P. *Religion and Science: The Basics,* Routledge, New York, 2012.

12. Dawkins, R. *The God Delusion*, Transword, London, 2006

13. Deacon, T. *The importance of what's missing, New Scientist,* 25 November 2011. http://www.sciencedirect.com/science/article/pii/S0262407911629187

14. Dunne, B. and Jahn, R. *Information and Uncertainty in Remote Perception Research. Journal of Scientific Exploration, 17(2),* 2003. 207-241 http://www.scientificexploration.org/journal/jse_17_2_dunne.pdf

15. Einstein, A. *Religion and Science, New York Times Magazine,* 9 November 1930. http://www.sacred-texts.com/aor/einstein/einsci.htm

16. Feynman, R. *QED: The strange theory of light and matter.* Princeton University Press, 2006.

17. Ghosh, P. *Higgs boson 'glimpsed' at Large Hadron Collider* BBC, 13 December 2011 http://www.bbc.co.uk/news/science-environment-16153804

18. Greene, B. *The Fabric of the Cosmos (Space, Time, and the Texture of Reality)* Random House, 2004, and Television Series: NOVA *The Fabric of the Cosmos* DVD—*Acclaimed physicist Brian Greene reveals a mind-boggling reality beneath the surface of our everyday world*. November 2011. http://www.shoppbs.org/product/index.jsp?productId=11645531&cp=&sr=1&kw=fabric+of+the+cosmos&origkw=fabric+of+the+cosmos&parentPage=search

19. Hammersely, M. (2008) *Troubles with Triangulation*, In: Bergman, Manfred Mas ed. *Advances in Mixed Methods Research*. London: Sage, pp.22–36 http://oro.open.ac.uk/20474/

20. Hawking, S. and Mlodinow, L. *The Grand Design, New answers to the ultimate questions of life*, Bantam Press, 2010.

21. Honorton, C. and Ferrari, D. *Future telling: A meta-analysis of forced-choice precognition experiments, 1935-87, Journal of Parapsychology*, 1989 http://lfr.org/LFR/csl/library/HonortonFerrari.pdf

22. Jacobsen, M. *Complementary Research Methods,* University of Calgary, http://www.ucalgary.ca/~dmjacobs/phd/methods/

23. Jacques, V. et al. *Experimental Realisation of Wheeler's Delayed-Choice Gedanken Experiment*, Science Vol. 315. no. 5814, pp. 966–968. 2007. Preprint available at http://arxiv.org/abs/quant-ph/0610241v1

24. Kennedy, P. *How to combine multiple research methods: Practical Triangulation, Johnny Holland* 2009. http://johnnyholland.org/2009/08/practical-triangulation/

25. Kim, Y. et al. *A Delayed-Choice Quantum Eraser. Physical Review Letters,* 2000.

26. Kolata, G. *Study Says DNA's Power to Predict Illness Is Limited. The New York Times,* 2 April 2012 http://www.nytimes.com/2012/04/03/health/research/dnas-power-to-predict-is-limited-study-finds.html?_r=2

27. Lepore, F. *Dissecting Genius: Einstein's Brain and the Search for the Neural Basis of Intellect*, The Dana Foundation, 2001.

28. Libet, B. *Neurophysiology of Consciousness: Selected Papers and New Essays*, Birkhauser 1993 http://books.google.com.au/books?id=s5T-XiEyJDgC&dq=about+350+msec+before&source=gbs_navlinks_s

29. Lommel, P. *Consciousness Beyond Life: The Science of the Near-Death Experience* HarperOne 2010

30. Long, J. and Perry, P. *Evidence of the Afterlife: The Science of Near-Death Experiences* HarperOne, 2010, http://www.amazon.com/Evidence-Afterlife-Science-Near-Death-Experiences/dp/0061452556/ref=sr_1_1?ie=UTF8&s=books&qid=1259790382&sr=8-1

31. Norenzayan, A. *The God Issue: Religion is the key to civilisation*, New Scientist, 22 March 2012. http://www.newscientist.com/article/mg21328562.100-the-god-issue-religion-is-the-key-to-civilisation.html

32. Olsen, W. *Triangulation in Social Research: Qualitative and Quantitative Methods Can Really Be Mixed Developments in Sociology*, 2004, ed. M. Holborn, Ormskirk: Causeway Press. http://research.apc.org/images/5/54/Triangulation.pdf

33. Parry, R. *Timor church massacre found*, The Independent, 1999, http://www.independent.co.uk/news/world/timor-church-massacre-found-1128979.html

34. Pedersen, O. *Galileo's Religion. Proceedings of the Cracow Conference, The Galileo affair: A meeting of faith and science.* Cracow: Dordrecht, D. 1984.

35. Polkinghorne, J., and Beal, N. *Questions of Truth, Fifty-one Responses to Questions about God, Science, and Belief.* Westminster John Knox Press, Kentucky, 2009

36. Polkinghorne, J. *One World, the interaction of science and theology*, Templeton Press, USA, 2007.

37. Precht, R. *Who am I? and if so, how many?* Sribe, 2011.

38. Radin, D. *Supernormal, Science, Yoga, and the Evidence for Extraordinary Psychic Abilities*. Crown Publishing, 2013.

39. Radin, D. *Entangled Minds, Extrasensory Experiences in a Quantum Reality*, Paraview Pocket Books, 2006.

40. Radin, D. *Electrodermal Presentiments of Future Emotions. The Journal of Scientific Exploration* (2004) http://www.psiarcade.com/research/radinpresentiment2004.pdf

41. Rosenblum, B. and Kuttner, F. *The observer in the quantum experiment*. Department of Physics, University of California Santa Cruz. Foundations of Physics, 32 (8), 2002. 1273-1293, as explained in Radin, D. *Entangled Minds, Extrasensory Experiences in a Quantum Reality*, Paraview Pocket Books, 2006. pp 218-219.

42. Rhine, J. and Pratt, J. *A review of the Pearce-Pratt distance series of ESP tests. Journal of Parapsychology*, 18, 1954. pp165–77. http://psychicinvestigator.com/demo/ESPdoc.htm

43. Russell, P. The Primacy of Consciousness presentation, Physics of Consciousness Conference, Virginia, AV, November 4-7, 2004.

44. Schmidt, B. *Professor Brian Schmidt's guide to the universe highlights*. Highlights of a spellbinding tour of the cosmos by the 2011 Nobel Physics Laureate Professor Brian Schmidt of the Australian National University. In *Unravelling the greatest mystery—the universe*, Peter Spinks, 28 March 2012. *The Age, http://media.smh.com.au/technology/tech-talk/professor-brian-schmidts-guide-to-the-universe-highlights-3173247.html*

45. Sheldrake, R. *The Science Delusion; Freeing the Spirit of Enquiry*, Coronet, UK, 2012.

46. Shiga, D. *LHC hints at why we exist, New Scientist*, 26 November 2011.

47. Solomon, Grant and Jane, *The Scole Experiment; Scientific Evidence for Life after Death*, UK, Judy Piatkus Publishers, 1999.

48. Surowiecki, J. *The Wisdom of Crowds*. Anchor (August 16, 2005) http://www.amazon.com/gp/product/0385721706/ref=as_li_qf_sp_asin_tl?ie=UTF8&tag=juanreyero-20&linkCode=as2&camp=1789&creative=9325&creativeASIN=0385721706

49. Surowiecki, J. *Mass Intelligence,* Forbes.com 2004 http://www.forbes.com/global/2004/0524/019.html

50. Talbot, M. *The Holographic Universe*, Harper Perennial, 1992, reissued 2011.

51. Targ, R. *The Reality of ESP: A Physicist's Proof of Psychic Abilities*, Quest Books, 2012.

52. Tart, Charles T. *The End of Materialism*. New Harbinger Publications, Inc., 2009.

53. Vincent J. et al. *Experimental Realisation of Wheeler's Delayed-Choice Gedanken Experiment, Science* Vol. 315. no. 5814, pp. 966–968 (2007). Preprint available at http://arxiv.org/abs/quant-ph/0610241v1

54. Wheeler, J. and Ford, K. *Geons, Black Holes, and Quantum Foam: A Life in Physics* (New York: Norton, 1998); Wheeler, J., *At Home in the Universe* (New York: American Institute of Physics, 1994), as discussed in Clayton, P. *Religion and Science: The Basics*, Routledge, New York, 2012.

55. Wheeler, J. and Ford, K. *Geons, Black Holes, and Quantum Foam: A Life in Physics* (New York: Norton, 1998); Wheeler, J., *At Home in the Universe* (New York: American Institute of Physics, 1994)

56. Zammit, Z. *A Lawyer Presents the Case for the Afterlife; Irrefutable Objective Evidence*. Gammell, 2006

57. Zammit, Z & Zammit, W. *A Lawyer Presents the Case for the Afterlife*. White Crow Books, 2013

58. *Prof Anton Zeilinger speaks on quantum physics*. at UCT Nov 2, 2011 http://www.youtube.com/watch?v=s3ZPWW5NOrw

NOTES

INTRODUCTION

1. Maslow, A. *The Psychology of Science: A Reconnaissance*, 1966, Ch. 2, p. 15.

2. Albert Einstein Archives, Hebrew University of Jerusalem, *http://www.albert-einstein.org/archives13.html*

CHAPTER 1

1. *http://edition.cnn.com/2013/04/18/us/planet-discovery*

2. 2005 movie, *How William Shatner Changed the World*. And *http://www.space.com/9705-top-10-star-trek-technologies.html*

3. *George Lucas predicts the future as NASA discovers planet orbiting two suns*, News.com.au, September 16, 2011. *http://www.news.com.au/technology/science/george-lucas-predicts-the-future-as-nasa-discovers-planet-orbiting-two-suns/story-fn5fsgyc-1226138632911*

4. *Deductive Reasoning Versus Inductive Reasoning*

    By Ashley Crossman *http://sociology.about.com/od/Research/a/Deductive-Reasoning-Versus-Inductive-Reasoning.htm*

5. Olsen, 2004, p.23.

6. Kennedy, 2009.

7. Ersberger and Kelle, 2003, p.461.

8. Surowiecki, 2004.

9. Ibid.

10. Ibid.

11. Ibid.

12. Jacobson, M. *Complementary Research Methods,* University of Calgary, *http://www.ucalgary.ca/~dmjacobs/phd/methods/*

13. *'Scientists only understand 4% of universe'* Published 29 July 2012 10:29 *http://rt.com/news/universe-physics-laws-energy-329/*

14. *http://science.howstuffworks.com/dictionary/astronomy-terms/question221.htm*

15. Attributed to Socrates by Plato, according to William L. Patty and Louise S. Johnson, *Personality and Adjustment,* p. 277 (1953). *http://www.bartleby.com/73/195.html*

16. *The double-slit experiment,* September 2002 issue of *Physics World* (p. 15). *http://physicsworld.com/cws/article/print/9745*

17. *http://amasci.com/weird/vindac.html http://amasci.com/weird/vindac.html#j16* and *http://www.scienceforthepublic.org/they-didnt-believe-it/A-Z#photon*

18. *http://ncu9nc.blogspot.com.au/2013/04/a-history-of-scientific-discoveries.html*

19. *Fire from the Sky. Science Debunked Meteorites,* Mysterious New Zealand, Tue, 29 Nov 2011 *http://www.sott.net/article/238325-Science-Debunked-Meteorites*

20. Ibid.

21. Wallace, A. *Notes on the Growth of Opinion as to Obscure Psychical Phenomena during the Last Fifty Years* The Alfred Russel Wallace Page hosted by Western Kentucky University.

    For a listing of Wallace's accomplishments see *http://en.wikipedia.org/wiki/Alfred_Russel_Wallace.*

22. For information on paradigm shift and Thomas Kuhn
    http://en.wikipedia.org/wiki/Paradigm_shift

23. Sinclair, U. *I, Candidate for Governor: And How I Got Licked* (1935), ISBN 0-520-08198-6; repr. University of California Press, 1994, p. 109. Also used by Al Gore, in *An Inconvenient Truth*, 2006.

24. Bell, E. *Mathematics, Queen and Servant of the Sciences* (1952) by https://archive.org/stream/MathematicsQueenAndServantOfScience/Bell-MathematicsQueenAndServantOfScience#page/n33/mode/1up/search/einstein

25. *Redesign My Brain*, Episode 2, 'Make Me Creative', aired 17 October 2013, ABC http://www.abc.net.au/tv/programs/redesign-my-brain-with-todd-sampson/#episode/DO1125H002S00

26. http://www.cracked.com/article_19140_8-movie-special-effects-you-wont-believe-arent-cgi.html

27. As reported by Lehrer, J. 'Why smart people are stupid', *The New Yorker*, June 2012 http://www.newyorker.com/online/blogs/frontal-cortex/2012/06/daniel-kahneman-bias-studies.html

28. Burkeman, O. "Daniel Kahneman: 'We're beautiful devices'" The Guardian UK, Tuesday 15 November, 2011. "http://www.theguardian.com/science/2011/nov/14/daniel-kahneman-psychologist"

29. West, R. et al. 'Cognitive sophistication does not attenuate the bias blind spot'. *Journal of Personality and Social Psychology* 2012 Sep. 103(3):506-19. doi: 10.1037/a0028857. Epub 2012 Jun 4. http://www.ncbi.nlm.nih.gov/pubmed?term=west%20stanovich%20meserve

30. http://www.liberalarts.wabash.edu/ncs/

31. As reported by Lehrer, J. 'Why smart people are stupid', *The New Yorker*, June 2012 http://www.newyorker.com/online/blogs/frontal-cortex/2012/06/daniel-kahneman-bias-studies.html

32. Ibid.

33. Dr David Halpern, Nudge in Government. *International Behavioural Insights Conference*, 2-3 June 2014, The United States Studies Centre, the University of Sydney, held at the Four Seasons Hotel, Sydney. http://bx2014.org/videos/nudging-in-government

34. Sheldrake, 2012, p. 84

35. Hawking, S. *Space and Time Warps*, Lecture, 1999. http://www.hawking.org.uk/space-and-time-warps.html

36. Sheldrake, 2012, pp. 88-108.

37. Hambling, D. *NASA validates 'impossible' space drive.* WIRED.CO.UK, 31 July 2014. http://www.wired.co.uk/news/archive/2014-07/31/nasa-validates-impossible-space-drive

38. Vincent, J. *NASA approves 'impossible' space engine design that apparently violates the laws of physics and could revolutionise space travel. The Independent.* 4 August 2014. http://www.independent.co.uk/news/science/nasa-approves-impossible-space-engine-design-that-apparently-violates-the-laws-of-physics-and-could-revolutionise-space-travel-9646865.html

39. Whittington, M. *NASA successfully tests engine that uses no fuel, violates the laws of physics,* Examiner.com. 1 August 2014. http://www.examiner.com/article/nasa-successfully-tests-engine-that-uses-no-fuel-violates-the-laws-of-physics

40. *Oral History Transcript — Dr John F. Clauser,* Interview with Dr John F. Clauser, By Joan Bromberg. At Walnut Creek, California, May 20, 2002. http://www.aip.org/history/ohilist/25096.html

41. http://www.youtube.com/watch?v=JKHUaNAxsTg

42. Dr Rupert Sheldrake talks about his banned TED talk on Skeptiko. http://www.youtube.com/watch?v=kAuxXvNVhgA

43. *The Ban on Rupert Sheldrake's TED Talk* 22 May, 2013
http://www.tricycle.com/blog/ban-rupert-sheldrakes-ted-talk

44. Ibid.

45. Cooper, D. *Deep-sea fish found to be warm-blooded* ABC Science, 15 May 2015 http://www.abc.net.au/science/articles/2015/05/15/4235761.htm

CHAPTER 2

1. American Heritage Dictionary http://dictionary.reference.com/browse/physics?s=t

2. Radin, 2013, p. 116.

3. Sheldrake, 2012, pp.213-4.

4. Episode 4: 'Are There More than Three Dimensions?' 29 June 2011, in *Through the Wormhole with Morgan Freeman*, Discovery Channel.

5. Hawking, S. and Mlodinow, L. *The Grand Design, New answer to the ultimate questions of life*, Bantam Press, 2010. p. 74.

6. Feynman, R. 2006 and Sheldrake 2012, discussion of Feynman's research pp. 60-61.

7. Walia, A. *The Illusion Of Matter: Our Physical Reality Isn't Really Physical At All*, December 5, 2013 http://www.collective-evolution.com/2013/12/05/the-illusion-of-matter-our-physical-material-world-isnt-really-physical-at-all/

8. Schmidt, B. 2011 video presentation http://media.smh.com.au/technology/tech-talk/professor-brian-schmidts-guide-to-the-universe-highlights-3173247.html also dark energy comprises the majority of the universe but no one knows what it consists of—Polkinghorne 2009, p.105.

9. Dark energy comprises the majority of the universe but no one knows what it consists of—Polkinghorne 2009, p.105.

10. Episode 4: 'Are There More than Three Dimensions?' 29 June 2011, in *Through the Wormhole with Morgan Freeman*, Discovery Channel.

11. Ibid.

12. Clayton, 2012, p.94.

13. Who, above a certain—predictable—number coming together, behave as a single entity rather than a group of ants—behaviour also observed in other communal animals, like bees and termites.

14. Hawking, S. and Mlodinow, L. 2010. pp 67-68.

15. Green, 2011.

16. Einstein, A., Podolsky, B., Rosen, N. 'Can Quantum-Mechanical Description of Physical Reality Be Considered Complete?' *Physical Review* 47 (10): (1935). pp.777–780.

17. Yin, J. et al. *Bounding the speed of 'spooky action at a distance'* Pan1

    1 Shanghai Branch, National Laboratory for Physical Sciences at Microscale,

    and Department of Modern Physics, University of Science and Technology of China, Shanghai 201315, China *http://arxiv.org/pdf/1303.0614v1.pdf*

18. *John Bell and the most profound discovery of science* PhysicsWorld.com, 1 December 1998 *http://physicsworld.com/cws/article/print/1998/dec/01/john-bell-and-the-most-profound-discovery-of-science*

19. Ursin, R. et al . 'Entanglement-based quantum communication over 144 km', *Nature Physics*, 3, 481–486 (2007).

20. Ibid..

21. Schmidt, B. 2011 video presentation *http://media.smh.com.au/technology/tech-talk/professor-brian-schmidts-guide-to-the-universe-highlights-3173247.html* also *dark energy comprises the majority of the universe but no one knows what it consists of*—Polkinghorne 2009, p. 105.

22. Greene, 2011.

23. Greene, 2011.

24. Letter from Einstein to the family of his lifelong friend Michele Besso, after learning of his death (March 1955), as quoted in *Science*

and the Search for God: Disturbing the Universe (1979) by Freeman Dyson, Ch. 17, 'A Distant Mirror' ; also quoted at *Einstein's God* (NPR)

25. Green, 2011.

26. Ibid.

27. Ibid.

28. Ibid.

29. As quoted in *What Life Means to Einstein, an Interview by George Sylvester Viereck*, in *The Evening Post*, 26 October 26, 1929 http://www.saturdayeveningpost.com/wp-content/uploads/satevepost/what_life_means_to_einstein.pdf.

30. Hawking, S. *Space and Time Warps*, Lecture, 1999. http://www.hawking.org.uk/space-and-time-warps.html

31. 'This says that certain quantities, like the position and speed of a particle, can't both have well-defined values. The more accurately the position of a particle is defined, the greater is the uncertainty in its speed, and vice versa. The uncertainty principle also applies to fields, like the electromagnetic field, or the gravitational field. It implies that these fields can't be exactly zeroed, even in what we think of as empty space. For if they were exactly zero, their values would have both a well-defined position at zero, and a well-defined speed, which was also zero. This would be a violation of the uncertainty principle. Instead, the fields would have to have a certain minimum amount of fluctuations.' Hawking, 1999.

32. Nathan, as published in *"Eagleworks, NASA's work on Warp Drive Revealed"*, 13 April 2013, *http://www.softmachine.net/2013/04/eagleworks-nasas-work-on-warp-drive-revealed/* Also see TODAY, August 2013, *NASA developing Star Trek-like warp drive for future space travel*

    http://www.todayonline.com/daily-focus/science/nasa-developing-star-trek-warp-drive-future-space-travel

33. See: *Edinburgh Review, Jan 1803, pp. 450-56*. The author, Henry Brougham (1778-1868), was a barrister who later became Lord

Chancellor of England. Brougham was an ardent advocate of the corpuscular theory of light. He savagely attacks Young's wave theory and Young himself. *http://homepages.wmich.edu/~mcgrew/brougham.htm*, Referenced in Lommel, 2010, p.216.

34. 'The double-slit experiment', September 2002 issue of *Physics World* (p15). http://physicsworld.com/cws/article/print/9745

35. *http://abyss.uoregon.edu/~js/21st_century_science/lectures/lec13.html*

36. Arndt, M. et al. *'Wave-particle duality of C60 molecules,'* Nature, London. pp. 401, 680– 682, 1999.

37. Feynman, R. 2006.

38. Rosenblum, B., and Kuttner, F. *'The observer in the quantum experiment'*. Department of physics, University of California Santa Cruz. *Foundations of Physics*, 32 (8), 2002. 1273-1293, as explained in Radin, D. *Entangled Minds, Extrasensory Experiences in a Quantum Reality*, Paraview Pocket Books, 2006. pp 218-219. And Adler 2011, p.46-7 citation of Niels Bohr's observation that our observation of the wave function causes it to collapse and a particle to appear at a particular time and place.

39. Jacques, V. et al. 2007.

40. Kim, Y. et al. 2000.

41. Jacques, V. et al. 'Experimental Realisation of Wheeler's Delayed-Choice Gedanken Experiment'. *Science* 315 (5814): pp. 966–968. *http://www.sciencemag.org/content/315/5814/966.full*

42. Kim, Y. et al. 'A Delayed Choice Quantum Eraser'. *Physical Review Letters* (2000). 84: 1–5. *http://www.bottomlayer.com/bottom/kim-scully/kim-scully-web.htm*

43. Hawking, S. and Mlodinow, L. 2010. pp. 82-140.

44. Cited in Radin, 2013, p. 240.

45. Adler 2011, p. 45.

46. '... the origin of the eukaryotic cell (our kind of cell, with mitochondria, which are not present in bacteria) was an even more momentous, difficult and statistically improbably step than the origin of life. The origin of consciousness might be another major gap whose bridge was of the same order of improbability' Dawkins 2006. p. 168. Citing Mark Radley—*The Cooperative Gene*.

47. Hawking, S. and Mlodinow, L. 2010. p. 162.

48. Dawkins, 2006. Discusses the existence of the impossible, including places 'inhabited by dragons' as inevitable due to the vastness of 'astronomical space', p. 419.

49. Adler, 2011, p. 44.

50. Adler, 2011, p. 47. Tegmark levels. Also, Nick Bostran, philosopher at the University of Oxford, doubts we will find evidence of these simulations, as the advanced civilisations are so smart.

51. Sir John Charlton Polkinghorne was Professor of Mathematical Physics at the University of Cambridge from 1968 to 1979, when he resigned his chair to study for the priesthood, becoming an ordained Anglican priest in 1982. He served as the president of Queens' College, Cambridge from 1988 until 1996. Polkinghorne is the author of five books on physics, and twenty-six on the relationship between science and religion—Polkinghorne, J., and Beal, N. *Questions of Truth, Fifty-one Responses to Questions about God, Science, and Belief.* Westminster John Knox Press, Kentucky, 2009, p. 105.

52. This was discussed already in the delayed-choice quantum eraser experiment. Kim, Y. et al. *A Delayed Choice Quantum Eraser* (2000), and supported by Hawkings et al *The Grand Design, New answer to the ultimate questions of life*, Bantam Press, 2010.

53. The reverse causality, i.e., an event later in time creating an event earlier in time, even billions of years is a popular theory, and one which is supported empirically—see Hawking, S. et al. *The Grand Design, New answer to the ultimate questions of life*, Bantam Press, 2010.

54. Wheeler, J. and Ford, K. 2012.

55. *The Library of Babel* (Spanish: *La biblioteca de Babel*) is a short story by Argentine author and librarian Jorge Luis Borges (1899–1986),

conceiving of a universe in the form of a vast library containing all possible 410-page books of a certain format.

56. Zeilinger, A. 2011.

CHAPTER 3

1. http://dictionary.reference.com/browse/biology?s=t

2. Soon, C. et al., 'Unconscious Determinants of Free Decisions in the Human Brain'. *Nature Neuroscience*, April 13th, 2008. Cited in *Brain Scans Can Reveal Your Decisions 7 Seconds before You 'Decide'*

   http://exploringthemind.com/the-mind/brain-scans-can-reveal-your-decisions-7-seconds-before-you-decide Also YouTube http://www.youtube.com/watch?v=N6S9OidmNZM#t=70

3. Precht, 2011, p. 51.

4. Sheldrake, 2012, p. 24. Famous mathematician and philosopher Bertrand Russell probably bests describes the naturalist's position in the following words: 'That man is the product of causes which had no prevision of the end they were achieving; that his origin, his growth, his hopes and fears, his loves and his beliefs, are but the outcome of accidental collocations of atoms; that no fire, no heroism, no intensity of thought and feeling, can preserve an individual life beyond the grave; that all the labours of the ages, all the devotion, all the inspiration, all the noonday brightness of human genius, are destined to extinction in the vast death of the solar system, and that the whole temple of Man's achievement must inevitably be buried beneath the debris of a universe in ruins—all these things, if not quite beyond dispute, are yet so nearly certain, that no philosophy which rejects them can hope to stand. Only within the scaffolding of these truths, only on the firm foundation of unyielding despair, can the soul's habitation henceforth be safely built.' In *A Free Man's Worship*, by Bertrand Russell, 1923, 6–7.

5. Cheery, K. *What Is Phrenology?* About.com—Psychology. http://psychology.about.com/od/historyofpsychology/f/phrenology.htm

6. Tart, C. *The End of Materialism*. New Harbinger Publications, Inc., 2009. pp. 19–33.

7. Hawking, S. and Mlodinow, L. 2010. pp 67–68.

8. Sheldrake, 2012, p. 84.

9. Green, 2011.

10. Sheldrake, 2012, p. 90.

11. Lommel, 2010, quoting philosopher David Chalmers, p. xv.

12. Russell, 2004.

13. Lommel, 2010. quoted p. xvi.

14. Hawking, S. and Mlodinow, L. 2010. pp. 82–140.

15. Polkinghorne, 2009. p. 119.

16. Wigner, E. *Symmetries and Reflections*. Cambridge, Mass: MIT Press, p. 171–184. As cited in Radin, 2006.

17. Cited in *The Purpose-Guided Universe: The God Theory* (Preface) http://www.thegodtheory.com/preface_pgu.htm

18. Zyga, L. *Free will is an illusion, biologist says* March 3, 2010. http://phys.org/news186830615.html

19. Wegner, D. *The Illusion of Conscious Will* MIT Press, 01/09/2003 and *Consciousness Explained* is a 1991 book by the American philosopher and vocal atheist Daniel Dennett which offers an account of how consciousness is an illusion which arises from interaction of physical and cognitive processes in the brain. He argues that in fact we are all 'zombies'.

20. ibid.

21. As cited in Lommel, 2010, pp. 203–4.

22. ibid.

23. Sheldrake, 2012, p. 213.

24. See Endnote 19.

25. Carey, B. 2008.

26. Sheldrake, 2012, p. 273. Also for a good summary go to *Is there scientific proof we can heal ourselves?* Rankin, L. MD at TEDx http://www.youtube.com/watch?v=LWQfe__fNbs

27. Cobb, L. et al. 'An evaluation of internal-mammary-artery ligation by a double-blind technic'. *New England Journal of Medicine*, 1959; 260 (22): 1115–8. http://www.nejm.org/doi/pdf/10.1056/NEJM195905282602204

28. Eippert, F. *et al.* 'Direct Evidence for Spinal Cord Involvement in Placebo Analgesia'. *Science*, 16 October 2009: Vol. 326 no. 5951 p. 404.

    http://www.sciencemag.org/content/326/5951/404.abstract?sid=943b1b88-b3f2-4551-9ea4-dcf93f16a05a

29. Hirschler, B. 'Placebo effect in the spine and mind', Friday, 16 October 2009. Reuters, *ABC Science*. http://www.abc.net.au/science/articles/2009/10/16/2715696.htm

30. Gever, J. *Placebos Advocated as Active Treatments*, Published: 20 February 2010 *MedPage Today* http://www.medpagetoday.com/PublicHealthPolicy/ClinicalTrials/18588?

31. Kaptchuk, T, et al. 'Components of placebo effect: randomised controlled trial in patients with irritable bowel syndrome'. *BMJ* 2008; 336 doi: http://dx.doi.org/10.1136/bmj.39524.439618.25 (Published 1 May 2008)

    Cite this as: BMJ 2008;336:999

    http://www.bmj.com/content/336/7651/999

32. Gever, J. 'Placebos Advocated as Active Treatments', Published 20 February 2010, *MedPage Today* http://www.medpagetoday.com/PublicHealthPolicy/ClinicalTrials/18588?

33. *Psychology Today* https://www.psychologytoday.com/conditions/dissociative-identity-disorder-multiple-personality-disorder

34. Braun, B. M.D., 'Treatment of Multiple Personality Disorder', *American Psychiatric Press*, 1986, as referenced in Goleman, D. 'Probing the

Enigma of Multiple Personality', *The New York Times* 25 June 1988, p. C1. *http://www.nytimes.com/1988/06/28/science/probing-the-enigma-of-multiple-personality.html?pagewanted=1*

35. Miller, S. et al. 'Optical differences in multiple personality disorder. A second look'. *Journal of Nervous Mental Disorders.* 1991 Mar; 179(3):132–5 *http://www.ncbi.nlm.nih.gov/pubmed/1997659*

36. Goleman, D. 'Probing the Enigma of Multiple Personality', *The New York Times* 25 June 1988, p. C1 *http://www.nytimes.com/1988/06/28/science/probing-the-enigma-of-multiple-personality.html?pagewanted=1*

37. Carey, B. 2012.

38. As quoted in Lommel, 2010, p. 198.

39. ibid.

40. Hawking, S. and Mlodinow, L. 2010. pp. 67–68.

41. Wigner, E. *Symmetries and Reflections.* Cambridge, Mass: MIT Press, p. 171–184. As cited in Radin 2006.

42. Physicist Heisenberg's (Heisenberg's uncertainty principle) view that quantum physics agrees with Plato that matter is ideas. Sheldrake 2012, p. 88.

43. Polkinghorne, J., and Beal, N. 2009, p. 122.

44. Polkinghorn 2009, p. 120 and Physicist Heisenberg's (Heisenberg's uncertainty principle) view that quantum physics agrees with Plato that matter is ideas. Sheldrake 2012, p. 88.

45. The *Flynn effect* is the name given to a substantial and long-sustained increase in intelligence test scores measured in many parts of the world (as discussed with other signs of *morphic resonance*, i.e., the phenomenon that an idea, once materialised, is received by people and other living things around the world)—Sheldrake 2012, pp. 206–209.

46. Sheldrake, R. *The presence of the past: morphic resonance and the habits of nature.* Icon Books. 2011.

http://books.google.com.au/
books?id=SyeKFT9hPTUC&pg=PT13&redir_esc=y#v=onepage&q&f=false

47. Cited in Radin, 2013, p. 313.

48. Cited in Radin, 2013, p. 313.

49. Sheldrake, 2012, p. 38.

50. As cited in Clayton, 2012, p. 78. For further information go to http://en.wikipedia.org/wiki/Orch-OR

51. Radin, 2013, p. 116

52. Sheldrake, 2012.

53. Posted by giblfiz 'People Who Have Virtually No Brains Are Living Among Us.' *Belligerati* | Civitatas Americanus Citius, Altius, Fortius 17 July 2006. http://www.belligerati.net/archives/2006/07/people_who_have.html

54. Feuillet, L. 'Brain of a white-collar worker', *The Lancet*, Volume 370, Issue 9583, p. 262, 21 July 2007.

55. Libet, 1993.

56. Larson, E. 2004, *Evolution: The Remarkable History of a Scientific Theory* Modern Library pp. 23–38.

57. Wallace, A. *Darwinism: An Exposition of the Theory of Natural Selection, with Some of Its Applications.* Macmillan. 1889. pp. 475–7.

58. Sheldrake, 2012.

59. Lommel, 2010, p. 264.

60. As referenced in Lommel, 2010, p. 276.

61. Kolata, 2012.

62. Sheldrake, R. *The presence of the past: morphic resonance and the habits of nature.* Icon Books. (2011).

http://books.google.com.au/
books?id=SyeKFT9hPTUC&pg=PT13&redir_esc=y#v=onepage&q&f=false

63. Jones, H. *http://www.amazon.co.uk/New-Science-Life-Rupert-Sheldrake/dp/1848310420*

64. Long, J. and Perry, P. *Evidence of the Afterlife: The Science of Near-Death Experiences* HarperOne, 2010. Dr Long asserts that there are nine arguments that prove the existence of life after death. These arguments have been generated through the study of consistencies from the hundreds of NDE accounts that he's complied over the years. Some of these arguments include: it can't be medically explained how people experience consciousness when they are clinically dead; blind people experience visual perceptions during their NDEs (even though, blind people do not dream in visuals); children give NDE details similar to those of adults, though they may have never been exposed to this concept; the 'life review's experience tend to reflect real events. These arguments, along with the others, are the primary basis for Long's proof assertion. Iris Green, 2010. *http://www.amazon.com/Evidence-Afterlife-Science-Near-Death-Experiences/dp/0061452556/ref=sr_1_1?ie=UTF8&s=books&qid=1259790382&sr=8-1* Also see Alexander, E. *Proof of Heaven: A Neurosurgeon's Journey into the Afterlife,* Simon and Schuster, 2012. *http://www.amazon.com/Proof-Heaven-Neurosurgeons-Journey-Afterlife/dp/1451695195*

65. Lommel, 2010, p. 279.

66. Balwin, G. et al., 'DNA Double Helices Recognise Mutual Sequence Homology in a Protein-Free environment', *Journal of Physical Chemistry* B 112, no. 4 (2008), pp. 1060–64. *http://www.ncbi.nlm.nih.gov/pubmed/18181611*

67. Lommel, 2010, p. 276.

68. Sheldrake, R. *The Presence of the Past* (London: Fontana, 1988) as referenced in Lommel, 2010.

69. 'Quantum biology: Do weird physics effects abound in nature,' By Jason Palmer and Alex Mansfield, BBC News and BBC Radio Science Unit, 28 January 2013 *http://www.bbc.co.uk/news/science-environment-21150047*

70. Palmer, J. and Mansfield, A. 'Quantum biology: Do weird physics effects abound in nature?' BBC News and BBC Radio Science Unit, 28 January 2013

    http://www.bbc.co.uk/news/science-environment-21150047

71. ibid.

72. As referenced in Lommel, 2010, p. 281.

73. http://skepdic.com/kirlian.html

74. NIH/National Human Genome Research Institute, 31 August 2005, 'Comparing the chimp and human genomes', based on a paper published in the 1 September 2005 issue of the journal *Nature*, the Chimpanzee Sequencing and Analysis Consortium described its landmark analysis which compared the genome of the chimp (*Pan troglodytes*) with that of human (*Homo sapiens*).
    http://genome.wellcome.ac.uk/doc_WTD020730.html

75. Gunter, D. & Dhand, R. 'The Mouse Genome', *Nature* 420, 509 (5 December 2002) doi:10.1038/420509a

    http://www.nature.com/nature/journal/v420/n6915/full/420509a.html

76. Radin, 2013, p.139.

77. See Endnote 81.

78. Martin, J. 'The Insanity Defence: A Closer Look'. Washingtonpost.com Staff Writer Friday, 27 February 1998. http://www.washingtonpost.com/wp-srv/local/longterm/aron/qa227.htm

79. Dawkins, R. *The God Delusion* 2006, for example, suggests that altruism is 'misfiring' within the brain—an accident, and then ironically refers to them as 'blessed mistakes' pp. 252–3.

80. *Darwin's Black Box: The Biochemical Challenge to Evolution*, Michael Behe, 1996, quoted in *Irreducible Complexity and Michael Behe* (retrieved 8 January 2006).

81. 'Why Intelligent Design Isn't Intelligent'. Review of: *Unintelligent Design*, by Mark Perakh; 2003; pp. 459. Prometheus Books (New York); ISBN: 1-5910-2084-0 http://www.ncbi.nlm.nih.gov/pmc/articles/PMC1103713/

82. However, the jury is still out on whether there is evidence that evolution happens gradually http://evolution.berkeley.edu/evolibrary/article/evo_51

83. Miller, K. *The Flagellum Unspun: The Collapse of 'Irreducible Complexity',* with reply *Still Spinning Just Fine: a response to Ken Miller* By William A. Dembski, 2.17.03, v. 1. http://designinference.com/documents/2003.02.Miller_Response.htm

84. Dawkins, 2006, pp. 162–180.

85. ibid. pp. 160–161.

86. Wallace, A. 1889. *Darwinism: An Exposition of the Theory of Natural Selection, with Some of Its Applications.* Macmillan. p. 477.

87. *The Selfish Gene* is a book on evolution by Richard Dawkins, published in 1976.

88. Polkinghorne, 2007, p.45. (and was dismissed by Popper as non-scientific in consequence).

89. Sheldrake, 2012, p. 221–222.

90. ibid. p. 222.

91. Radin, 2006, p. 195.

92. ibid. p.198.

93. Information provided by Dean Radin to the author on 9 November 2015.

94. For a good description of Benford's law and its applications go to http://en.wikipedia.org/wiki/Benford's_law

95. 'Indicators of Deception'. Forensics executive Director Rob Cockerell speaks with Rafael Epstein on ABC Radio 774, ABC 2011. Mel-

bourne. And Paul Francois & Enrique Garcia *Liar! Tips for Detecting Deception*, 2007 http://www.corrections.com/news/article?articleid=16247

96. Radin, D. *The Conscious Universe* 1997 HarperEdge, http://www.deanradin.com/Chapter1.html

97. http://www.ipcc.ch/pdf/assessment-report/ar4/wg1/ar4-wg1-chapter2.pdf

98. Radin, D. *Selected Peer-Reviewed Journal Publications on Psi Research*, 2013 http://noetic.org/research/psi-research/

99. Rhine, J. and Pratt, J. 1954.

100. Radin, D. *Men who stare at photons*, Electric Universe Conference, January 2013 http://www.youtube.com/watch?v=FMXqyf13HeM

101. *The Humanistic Psychologist*, 33(4) 293-303, 2005.

102. See Endnote 105.

103. Duane T.D. Behrendt T. 'Extrasensory electro-encephalographic induction between identical twins', *Science*, 1965:150:367. Targ R. Puthoff HE. 'Information transmission under conditions of sensory shielding', *Nature*, 1974; 251: pp. 602–607.

104. Archterberg, J. Ph.D, et al. 'Evidence for Correlations between Distant Intentionality and Brain Function in Recipients: A Functional Magnetic Resonance Imaging Analysis'. *The Journal of Alternative and Complementary Medicine*. Volume 11, Number 6, 2005, pp. 965–971. http://www.jeanneachterberg.com/achetal.pdf

105. Richards, T. et al. 'Replicable Functional Magnetic Resonance Imaging Evidence of Correlated Brain Signals Between Physically and Sensory Isolated Subjects'. *The Journal of Alternative and Complementary Medicine*. Volume: 11 Issue 6. 6 January 2006. http://online.liebertpub.com/doi/abs/10.1089/acm.2005.11.955

106. Dunne, B. and Jahn, R. 'Information and Uncertainty In Remote Perception Research'. *Journal of Scientific Exploration*, 17(2), (2003) pp. 207–241.

http://www.scientificexploration.org/journal/jse_17_2_dunne.pdf

107. Interview with Prof Targ—*Thinking Allowed*
http://www.youtube.com/watch?v=mHyVbYz16DM

108. Targ, 2012, http://www.amazon.com/The-Reality-ESP-Physicists-Abilities/dp/0835608840

109. ibid.

110. Schwartz, S. *Opening the infinite: The art and science of nonlocal awareness* Buda, Texas: Nemoseen Media, 2007.
http://www.thefreelibrary.com/
Opening+to+the+Infinite%3A+The+Art+and+Science+of+Nonlocal+Awareness.-a0176480173

111. Targ, 2012, http://www.amazon.com/The-Reality-ESP-Physicists-Abilities/dp/0835608840

112. Jahn, R. and Dunne, B. *Margins of Reality: The Role of Consciousness in the Physical World.* New York, Harcourt Brace Jovanovich, 1987, pp. 91–123. As cited in Talbot, 1992. p. 123.

113. Radin, 2013, p. 240.

114. ibid. p. 253.

115. ibid p. 264.

116. ibid. p. 265.

117. ibid. p.265.

118. Schmidt, H. and Schlitz,M. 'A large-scale pilot PK experiment with pre-recorded random events'. 1988. *Mind Science Foundation Research Report*, San Antonio, TX: Mind Science Foundation (See also abstract in Research in Parapsychology 1991).

119. Murphy, B. *The Grand Illusion: A Synthesis of Science and Spirituality*, Book One. 2012, BalboaPressAU. p. 81.

120. Radin, 2013, p. 264.

121. Radin, D. 2004.

122. National Public Radio's *Science Friday* program (May 1999) in Radin (2006, p. 170).

123. Radin, 2013, p. 164.

124. http://carlossalvarado.wordpress.com/2013/09/21/people-in-parapsychology-ii-julia-mossbridge/ Parapsychology. 2013.

125. Ma, X. et al., 'Experimental delayed-choice entanglement swapping', *Nature Physics* (2012).

126. Honorton, C. and Ferrari, D. 1989.

127. Tart, 2009. p. 105.

128. Banghui, W. 'Evidence of the Existence of Abnormal States of Matter', Defence Intelligence Agency translation, *Chinese Journal of Somatic Science*, First Issue, 36, 1990, and other research cited in Lommel, 2010, p. 306.

129. http://www.ehow.com/list_6186158_signs-show-before-natural-disaster.html

130. Sheldrake, R. and Smart, P. 'A Dog That Seems to Know When His Owner is Coming Home: Videotaped Experiments and Observations', *Journal of Scientific Exploration* 14, 233-255 (2000).

131. Streby, H. et al. 'Tornadic Storm Avoidance Behaviour in Breeding Songbirds'. *Current Biology* Published online 18 December 2014 http://www.cell.com/current-biology/abstract/S0960-9822(14)01428-6

132. Yang, S. *Sensing distant tornadoes, birds flew the coop. What tipped them off?* UC Berkeley New Centre. 18 December 2014. https://newscenter.berkeley.edu/2014/12/18/infrasound-as-early-storm-warning-for-birds/

133. Alvarez, F. 'Anticipatory alarm behaviour in Bengalese finches'. *Journal of Scientific Exploration* 2010;24(4): 599–610. http://www.scientificexploration.org/journal/full/jse_24_4_full.pdf#page=27 As cited in Radin, 2013, p.176

134. It has been researched in telephone call prediction in women producing milk let downs when away from their babies and in unique cases of people who seem to have been able to use such talents to compensate for severe disabilities (as in a case in which a disabled boy was found to be able to read through his mother's mind). (Sheldrake 2012, p.234-245).

135. Carter, 2012.

136. Radin, 2013. pp. 74-75.

137. ibid. p. 194.

138. http://noetic.org/research/psi-research/

139. IUn *The Universe in Light of Modern Physics*, first published in English in 1931. http://archive.org/stream/universeintheligo32967mbp#page/n11/mode/2up

140. See the work of Nobel Prize winner Charles Robert Richet, in his book *The Sixth Sense*, 1928 http://en.wikipedia.org/wiki/Charles_Richet

141. Tart, 2009, pp. 19–32.

142. See Endnote 67.

143. Lommel, 2010, p. 20.

144. Michael E. Tymn, VINE™ VOICE, 2010.

145. Lommel, 2010, p. 244.

146. ibid. pp. 17-43.

147. ibid. p. 156.

148. ibid. pp. 148–149.

149. Greyson (US) study 'Incidence and Correlates of Near-Death Experiences in a Cardiac Care Unit', *General Hospital Psychiatry* 25 (2003), p. 275, and Parnia et al. 'A Qualitative and Quantitative Study of the

Incidence, features and Aetiology of Near-Death Experiences in Cardiac Arrest Survivors', *Resuscitation* 48 (2001) p. 151.

150. Greyson, B. 'Incidence and Correlates of Near-Death Experiences Out of Their Bodies or Out of Their Minds?', *The Lancet* 355 (2000), pp. 460–463.

151. Lommel, 2010, p. 186.

152. ibid.

153. Cowey, A. and Walsh, V. Chapter 26: 'Tickling the brain: studying visual sensation, perception and cognition by transcranial magnetic stimulation'. In Casanova, Christian; Ptito, Maurice. *Vision: From Neurons to Cognition, Volume 1*. Gulf Professional Publishing. pp. 411–25. ISBN 9780444505866. (2001).

154. Beauregard, M., 'Mind Does Really Matter: Evidence for Neuroimaging Studies of Emotional Self-regulation, Psychotherapy, and Placebo Effect'. *Progress in Neurobiology* 81, no. 4 (2007), pp 218–36 as referenced in Lommel, 2010, p. 193.

155. Wallace, A. 'Notes on the Growth of Opinion as to Obscure Psychical Phenomena During the Last Fifty Years'. The Alfred Russel Wallace Page hosted by Western Kentucky University. For a listing of Wallace's accomplishments see http://en.wikipedia.org/wiki/Alfred_Russel_Wallace

156. Graber, C 'Snake Oil Salesmen Were on to Something'. *Scientific American*. 2007. Retrieved 2011-12-04. http://www.scientificamerican.com/article.cfm?id=snake-oil-salesmen-knew-something

157. Hutton, R. *The Triumph of the Moon: A History of Modern Pagan Witchcraft*. Oxford and New York: Oxford University Press. 1999. p. 342. http://en.wikipedia.org/wiki/Witch_trials_in_the_early_modern_period#cite_note-139

158. Levack, B. (*The Witch-Hunt in Early Modern Europe*) multiplied the number of known European witch trials by the average rate of conviction and execution, to arrive at a figure of around 60,000 deaths. Anne Lewellyn Barstow (*Witchcraze*) adjusted Levack's estimate to account for lost records, estimating 100,000 deaths. Ronald

Hutton (*Triumph of the Moon*) argues that Levack's estimate had already been adjusted for these, and revises the figure to approximately 40,000. *http://en.wikipedia.org/wiki/Witch_trials_in_the_early_modern_period#cite_note-7*

159. Zammit, 2013, p. 78.

160. *http://www.encyclopedia.com/doc/1G2-3403802097.html*

161. Zammit, 2006, p. 58–64.

162. Cited in Zammit, 2013, p. 101, *SPR Proceedings* Col 13, 1898, H 10.

163. Cited in Zammit, 2013, p. 101, Richet 1927.

164. Neher, A. *Paranormal and Transcendental Experience: A Psychological Examination.* Dover Publications. 2011. pp. 217–218. ISBN 978-0486261676.

165. – Zammit, 2006, p. 58–64.

166. Tymn, M. *Resurrecting Leonora Piper: How Science Discovered the Afterlife*, 2013, White Crow Books.

167. Solomon, 1999, p. 103.

168. ibid. pp. 160–61.

169. ibid. p. 16.

170. As cited in Zammit, 2013, p. 78. Gary E. Schwartz PhD., Deepak Chopra MD. *The Afterlife Experiments: Breakthrough Scientific Evidence of Life After Death*, Atria Books, 2002.

CHAPTER 4

1. Hildebrand, Ulrich. 1988. 'Das Universum—Hinweis auf Gott? in *Ethos* (die Zeitschrift für die ganze Familie), No. 10, Oktober. Berneck, Schweiz: Schwengeler Verlag AG. Reprinted by permission of the publisher, Schwengeler Verlag AG.

2. http://dictionary.reference.com/browse/Religion?s=t

3. 'Lectures XVI and XVII: Mysticism' in *The Varieties of Religious Experience*, by William James, cited in Radin, 2013. p. 51.

4. Rossano, M. 'The Religious Mind and the Evolution of Religion', in the *Review of General Psychology*, 2006, Vol. 10, No. 4, 345-364. p. 349.

5. ibid. p. 354.

6. Norenzayan, A. 'Religion is the key to civilisation', *New Scientist*, 22 March 2012.

7. Clayton, 2012, p. 48.

8. Sheldrake, 2012, p. 321.

9. Clayton, 2012, p. 82.

10. Radin, 2013. p. 4.

11. Parry, R. 'Timor church massacre found', *The Independent*, 1999, http://www.independent.co.uk/news/world/timor-church-massacre-found-1128979.html

12. Sheldrake, 2012, p. 278 and p. 338.

13. Smart, Simon. 'God's truth, believers are nicer', *The Sydney Morning Herald*, 9 September 2011.

    http://www.smh.com.au/opinion/society-and-culture/gods-truth-believers-are-nicer-20110908-1jzrl.html

14. Botton, 2012, p. 209.

15. ibid. p. 182.

16. Zwartz, B. 'The Search for Meaning', *The Sydney Morning Herald*, 21 December 2013, *News Review*. p. 1.

17. ibid..

18. Sheldrake, 2012, p. 24. Famous mathematician and philosopher Bertrand Russell probably bests describes the naturalist's position in the following words: 'That man is the product of causes which had no prevision of the end they were achieving; that his origin, his growth, his hopes and fears, his loves and his beliefs, are but the outcome of accidental collocations of atoms; that no fire, no heroism, no intensity of thought and feeling, can preserve an individual life beyond the grave; that all the labours of the ages, all the devotion, all the inspiration, all the noonday brightness of human genius, are destined to extinction in the vast death of the solar system, and that the whole temple of Man's achievement must inevitably be buried beneath the debris of a universe in ruins—all these things, if not quite beyond dispute, are yet so nearly certain, that no philosophy which rejects them can hope to stand. Only within the scaffolding of these truths, only on the firm foundation of unyielding despair, can the soul's habitation henceforth be safely built.'. In *A Free Man's Worship*, by Bertrand Russell, 1923, 6-7.

19. Michael Persinger of Laurentian University in Ontario claims to have created religious experiences in the brain through electrical charges from his *God helmet*, which generates weak electromagnetic fields and focuses them on particular regions of the brain's surface. Although a 2005 attempt by Swedish scientists to replicate Persinger's God helmet findings failed, researchers are not yet discounting the temporal lobe's role in some types of religious experience ... by David Biello From the October 2007 issue of *Scientific American Mind* from Scientific American website. *http://www.bibliotecapleyades.net/ciencia/ciencia_braino3.htm*

20. Cowey, A. and Walsh, V. Chapter 26: 'Tickling the brain: studying visual sensation, perception and cognition by transcranial magnetic stimulation'. In Casanova, Christian; Ptito, Maurice. *Vision: From Neurons to Cognition, Volume 1*. Gulf Professional Publishing. 2001. pp. 411–25. ISBN 9780444505866.

21. US National Academy of Science: *Religious Issues*: *http://www.nap.edu/openbook.php?record_id=5787&page=58*

22. 'Science without religion is lame, religion without science is blind ... It is cosmic religious feeling that gives a man such strength. A contemporary has said, not unjustly, that in this materialistic age of ours the serious scientific workers are the only profoundly religious peo-

23. 'I had no intention to write atheistically... I can see no reason, why a man, or other animal, may not have been aboriginally produced by other laws; & that all these laws may have been expressly designed by an omniscient Creator, who foresaw every future event & consequence. But the more I think the more bewildered I become.' Charles Darwin to Asa Gray, 22 May 1860, in *The Correspondence of Charles Darwin*. Vol. 8 (Cambridge University Press), p. 224. Also in the 6th Edition of *The Origin of Species*, exasperated by the overblown interpretations of his ideas, Darwin wrote. 'As my conclusions have lately been much misinterpreted ... may I be permitted to remark that in the first edition of the work, and subsequently, I placed in a most conspicuous position "I am convinced that natural selection has been the main, but not the exclusive, means of modification." This has been of no avail. Great is the power of steady misrepresentation.' Cited in Polkinghorne, 2009, p. 141.

24. Newton was as much a theologian as a physicist. Arguably he thought of himself more the former. In Newton's *Principia*, he concluded that humans know God only by examining the evidence of His creations: 'This most beautiful system of the sun, planets, and comets could only proceed from the counsel and dominion of an intelligent and powerful Being. He is eternal and infinite, omnipotent and omniscient; that is his duration reaches from eternity to eternity; his presence from infinity to infinity; he governs all things, and knows all things that are or can be done. We know him only by his most wise and excellent contrivances of things, and final causes; we admire him for his perfection; but we reverence and adore him on account of his dominion; for we adore him as his servants.'

25. Pedersen, O. 1984.

26. Planck, M. in his book, *Where is Science Going?* (1932), 1977 reprint, p. 168. http://crestroyertheory.com/2013/11/22/max-planck-quotes/

27. http://www.abc.net.au/tv/qanda/txt/s3469101.htm Dawkins vs Cardinal George Pell on *Q&A* ABC.

28. Radin, 2013, p. 36.

29. Zwartz, B. 'The Search for Meaning', *The Sydney Morning Herald*, 21 December 2013, *News Review* p. 4.

30. Gilbert Keith Chesterton, Chesterton Society, http://www.chesterton.org/, , http://alanarchibald.homestead.com/ChestertonQuotes.html

31. Clayton, 2012. p.158-9.

32. See 'Reality in Buddhism' in Wikipedia: https://en.wikipedia.org/wiki/Reality_in_Buddhism

33. Clayton, 2012. p.57.

34. Robinet, I. *Taoism: Growth of a Religion* (Stanford: Stanford University Press, 1997 [original French 1992]). ISBN 0-8047-2839-9.

35. By AG staff with AAP, 'DNA confirms Aboriginal culture one of Earth's oldest', *Australian Geographic*, September 2011. http://www.australiangeographic.com.au/news/2011/09/dna-confirms-aboriginal-culture-one-of-earths-oldest/

36. Ma, X. et al. 'Experimental delayed-choice entanglement swapping', *Nature Physics* (2012)

37. Radin, 2013. p. 108.

38. ibid. p. 61.

39. ibid. p. 28.

40. ibid. p. 123.

41. ibid. p. 125.

42. Interview with Prof Targ—*Thinking Allowed* http://www.youtube.com/watch?v=mHyVbYz16DM

43. Talbot, 1992, p. 110.

44. *Mystics of the Church, Therese Neumann—Mystic Victim Soul & Stigmatic*. http://www.mysticsofthechurch.com/2009/12/therese-neumann-mystic-victim-soul.html#sthash.yvjkxaxG.dpuf

Sourced: Vogl, A. *Therese Neumann—Mystic and Stigmatist*, Tan Books, 1987, and also Schimberg, A. *The Story of Therese Neumann*, Bruce Publishing Co., 1947.

45. ibid.

46. As cited in Radin, 2013 p. 29.

47. Planck M. 'Consciousness matters'. *The Observer*, 25 January 25 1931, as cited in Radin, 2013, p. 311.

48. Goswami, Amit. *The Self-Aware Universe*, Tarcher, 1995. p. 8.

49. As cited in Radin, 2013, pp. 29, 54, 57.

50. Durkheim, Émile. *The Elementary Forms of Religious Life*. Translation by Joseph Ward Swain. New York: The Free Press. 1965. p. 454, 249–50. *https://archive.org/details/elementaryformsooodurkrich*

51. Buehler, A. *The Twenty-first-century Study of Collective Effervescence: Expanding the Context of Fieldwork*—Vol 7, No 1 (2012)—*Fieldwork in Religion*. p. 75.

    http://www.academia.edu/2614103/The_Twenty-first-century_Study_of_Collective_Effervescence_Expanding_the_Context_of_Fieldwork

52. Buehler, A.—*Fieldwork in Religion*. 2012 pp. 76 and 78.

53. De Santillana, G. *The Crime of Galileo*. Chicago, IL: University of Chicago Press. 1978. 8.

54. Buehler, A. *Fieldwork in Religion*. 2012 p. 94.

55. Lemonick, M. '"God Particle" Found? Search for the Higgs Boson Narrows', *Time Science*, Tuesday, 13 December 2011. http://www.time.com/time/health/article/0,8599,2102190,00.html

56. Shiga, D. 'LHC hints at why we exist', *New Scientist*, 26 November 2011.

57. De Botton, 2012, pp. 202-03.

58. American Dr Stuart Hameroff and British physicist Sir Roger Penrose developed a quantum theory of consciousness asserting that our souls are contained inside structures called microtubules which live within our brain cells. 'Scientists offer quantum theory of soul's existence'. 31 October, 2012. http://www.news.com.au/lifestyle/quantum-scientists-offer-proof-soul-exists/story-fneszs56-1226507452687

59. This was discussed already in the delayed-choice quantum eraser experiment. Kim, Y. et al. *A Delayed Choice Quantum Eraser*, 2000, and supported by Hawking et al., *The Grand Design, New answer to the ultimate questions of life*, Bantam Press, 2010.

60. The reverse causality, i.e., an event later in time creating an event earlier in time—even billions of years—is a popular theory, and one which is supported empirically. See Hawking, S. et al., *The Grand Design, New answer to the ultimate questions of life*, Bantam Press, 2010.

61. Wheeler, J. and Ford, K. *Geons, Black Holes, and Quantum Foam: A Life in Physics* (New York: Norton, 1998); Wheeler, J., *At Home in the Universe* (New York: American Institute of Physics, 1994), as discussed in Clayton, P. *Religion and Science: The Basics*, Routledge, New York, 2012.

62. Planck, M. 'Consciousness matters'. *The Observer*, 25 January 1931, as cited in Radin, 2013, p. 311.

63. Linde, A. Department of Physics, Stanford University, Stanford, CA 94305, USA. 'Inflation, Quantum Cosmology and the Anthropic Principle', to appear in *Science and Ultimate Reality: From Quantum to Cosmos*, honouring John Wheeler's 90th birthday. JD Barrow, PCW Davies, & CL Harper eds. Cambridge University Press (2003) http://arxiv.org/pdf/hep-th/0211048.pdf

64. De Botton, 2012, p. 312.

CHAPTER 5

1. Wallace, D. 'Plain old untrendy troubles and emotions', *The Guardian*, 20 September 2008. http://www.theguardian.com/books/2008/sep/20/fiction

2. Atkins, E. M. D. 'Coping with Multiple Sclerosis—Visual Problems in MS-Part 1'. *Optic Neuritis*, 2009 *http://www.msfocus.org/article-details.aspx?articleID=380*

3. Cycleback, D. *A Look at How Humans Think and See*, 2005, http://www.cycleback.com/conceits.html

4. Pribram, K. 'The Neurophysiology of Remembering', *Scientific American* 220, January 1969, p. 76-78 as referenced in Talbot, 1992, p. 163—tests on monkeys.

5. Talbot, M. 1992, p. 163.

6. *http://en.memory-alpha.org/wiki/The_Traveler*

7. Using subspace to travel to other places in the universe instantly would present huge risks as well as opportunities, for us and for those whom we might discover. It is perhaps of little surprise that the space-time distance between us and any other potential life-sustaining solar systems is thousands of light years. It provides a natural barrier which cannot be crossed until a significant and game-changing advance is made. It protects all parties until such a turning point is considered permissible.

8. Hawking, S. and Mlodinow, L. 2010. pp. 82-140.

9. As quoted in Dawkins, 2006, p. 411.

10. Pollack, G.—*The Ills of Science*—Electric Universe 2013 Conference. *http://www.youtube.com/watch?v=TzobC_4_xLo*

11. ibid.

12. Institute for Venture Science *http://www.theinstituteforventurescience.net/*

13. Feyerabend, P. *Science in a free society*, 1982. New Left Books, London.

14. Friesike, S. et al. *Opening science: towards an agenda of open science in academia and industry*, Springerlink.com, 25 November 2014.

15. Nielsen, M. *An informal definition of open science*. The Open Science Project. (2011). *http://www.openscience.org/blog/?p=454*

16. Friesike, S. et al. *Opening science: towards an agenda of open science in academia and industry*, Springerlink.com, 25 November 2014.

17. Stephan A. Schwartz is the Senior Samueli Fellow for Brain, Mind and Healing of the Samueli Institute, and a research associate of the Cognitive Sciences Laboratory of the Laboratories for Fundamental Research. He is the columnist for the journal *Explore*, and editor of the daily web publication Schwartzreport.net , both of which cover trends that are affecting the future. He also writes regularly for *Huffington Post*. Previously he was the founder and research director of the Mobius Laboratory, and director of research at the Rhine Research Centre, senior fellow of The Philosophical Research Society, special assistant for research and analysis to the Chief of Naval Operations, and an editorial staff member of *National Geographic*. For 40 years he has been studying the nature of consciousness, particularly that aspect independent of space and time.

18. Stephan A Schwartz*: Non-Local Consciousness and Exceptional Experiences* Published on 4 March 2014. *A Transformation Dialogue with Stephan A Schwartz and Dr Craig Weiner. http://www.youtube.com/watch?v=VWw2B5Y7tG0*

19. ibid.

20. ibid.,.

21. Talbot 1992, p. 49.

22. Ullman, M. 'Wholeness and Dreaming', in *Quantum Implications*, Edited by Basil, J. Hiley, B. and Peat. D, Routledge & Kegan Paul, 1987. p. 393, as referenced in Talbot 1992, p. 63. *http://siivola.org/monte/papers_grouped/copyrighted/Dreams/Wholeness_and_dreaming.htm*

23. Zimmerman, M. 'Buddhists Join the Clergy Letter Project and Call for the Teaching of Evolution', *Huffington Post,* 29 August, 2012. *http://www.huffingtonpost.com/michael-zimmerman/buddhists-join-the-clergy-letter-project-and-call-for-the-teaching-of-evolution_b_1830935.html*

24. *Is Alien Life Out There?* Vatican Observatory Co-Hosts Science Conference in Arizona, Space.com, *http://www.space.com/25060-vatican-observatory-alien-life-conference.html*, The Vatican Observatory. *http://www.vaticanobservatory.va/content/specolavaticana/en.html*

25. Personal conversations with Arthur Buehler, senior lecturer in Religious Studies at Victoria University, New Zealand.

26. Pribram, K. 'The Neurophysiology of Remembering', *Scientific American* 220, January 1969, p. 75 as referenced in Talbot, 1992.

27. Pribram, K. *Languages of the Brain*, Monterey, California, Wadsorth Publishing, 1977, p. 169, as referenced in Talbot, 1992, p. 25.

28. Talbot, 1992. p. 158—interviews with William Tiller.

29. Esbjorn-Hargens, S. *An All-Inclusive Framework for the 21st Century, An Overview of Integral Theory*, Integral Post, Integral Life, 2009, *https://integrallife.com/integral-post/overview-integral-theory*

30. Ken Wilber on *The Good, The True, and The Beautiful part 1*
https://www.youtube.com/watch?v=WjFxGvTqpTA

31. See Endnote 29.

32. Wilber, K. A *Theory of Everything: An Integral Vision for Business, Politics, Science, and Spirituality*, 2000, p. xi—Notes to the Reader.

33. *A Brief History of Integral with Ken Wilber https://www.youtube.com/watch?v=i2dUo6hVOsk*

34. Integral Communication The Philosopher Kings
https://www.youtube.com/watch?v=CxZYzxylQvI&list=PL66bRS2DhVi_LE2vyM1zFDz_xC2HqaPoZ

35. 'Less than 10 per cent of smokers who attend an Easyway clinic ask for a refund of their £240 under the three-month guarantee, but these days the organisation still makes only the more sober boast of a 50 per cent success rate—a figure corroborated by two studies published in peer-reviewed journals. And it's worth noting that a success rate of 50 per cent after 12 months would be a significant, two- or threefold improvement on the successes currently reported by NHS smoking-

NOTES

cessation services.' Laurence Phelan, 'How Allen Carr saved my lungs', *The Independent*, 2012. http://www.independent.co.uk/life-style/health-and-families/features/how-allen-carr-saved-my-lungs-7562758.html

36. Petty, M., *Daily Mail* 2006, http://www.dailymail.co.uk/health/article-399517/Quit-smoking-guru-Allen-Carr-I-months-live.html

37. Phelan, L. 'How Allen Carr saved my lungs', *The Independent*, 2012. http://www.independent.co.uk/life-style/health-and-families/features/how-allen-carr-saved-my-lungs-7562758.html

38. McCoy, T. 'Meet the outsider who accidentally solved chronic homelessness', *The Washington Post*, 3 May 2015. http://www.washingtonpost.com/news/inspired-life/wp/2015/05/06/meet-the-outsider-who-accidentally-solved-chronic-homelessness/

39. ibid.

APPENDIX

1. *A Delayed-Choice Quantum Eraser* by Kim, Y. et al. http://xxx.lanl.gov/pdf/quant-ph/9903047 (citations omitted) Phys.Rev.Lett. 84 1-5 (2000). *http://www.bottomlayer.com/bottom/kim-scully/kim-scully-web.htm*

2. http://www.bottomlayer.com/bottom/kim-scully/kim-scully-web.htm
   Ross Rhodes, RhodesR@BottomLayer.com

INDEX

Aboriginal, Australian, 185, 194
Achterberg, Jeanne, 131
Adams, Douglas, 197
Aharonov, Yakir, 187
Allen, Paul, 140
Altshuller, Genrich, 213
Alvarez, Fernando, 145
Ames Room, 22
Anthropic Principle, 63
Apports, 160
Bach, Johann Sebastian, 77, 212, 213,
Backster, Cleve, 112
Bazerman, Max, 26
Beethoven, Ludwig van, 77
Behaviour Insights Team, 26
Behe, Michael, 116
Bell, Eric Temple, 21
Bell's Theorem, 45
Benford's law, 125
Benson, Herbert, 188
Berkeley, George, 62
Besso, Michele, 48
Bias blind spot, 25
Bierman, Dirk, 141
Big Bang, 16, 41, 49, 78, 79, 82, 88, 89, 105, 114, 120, 195, 215
Black Holes, 16, 100, 226
Blackmore, Susan, 86
Bohm, David, 185, 211-230
Bohnet, Iris, 26
Bohr, Niels, 10, 40, 61, 62, 192
Borges, Jorge Luis, 68

Botton, Alain de, 173, 198, 201
Brahman, The supreme spirit, 180
Brain Elasticity, 96
Brain Size, 103
Braun, Bennet, 95
Breatharianism, 188
Buckyballs, 55
Buddha, 6, 179, 223
Buddhism, 172, 178, 179, 187, 188, 221
Buehler, Arthur, 194-223
Cannae Drive, 29
Carr, Allen, 239-242
Casimir effect, 51
Chladni, Ernst, 18
Christianity, 178, 183, 189, 223
Churchill, Winston, 157
Citizen Science, 218
Clairvoyance, 127, 147, 165, 187
Clauser, John F, 30
Clayton, Philip, 172, 179
Clergy Letter Project,The, 221
Cogdell, Richard, 112,
Collective Effervescence, 193-201
Consciousness, 83-167, 179, 192-235
d'Espagnat, Bernard, 62, 135
Dalai Lama, 172, 180, 192, 193, 200, 221
Dark energy, 14
Dark matter, 16, 40, 41, 113
Darwin, Charles, 19, 79, 105, 106, 119, 177

Dawkins, Richard, 63, 116
De Gaulle, General Charles, 157
DeFreeze, Donald, 134
Degado-Romero, Edward, 130
Delayed Choice Experiments, 56, 57, 66, 137
Delayed Choice Quantum Eraser, 56, 60, 83, 137,
Dennett, Daniel, 86
Descartes, Rene, 84, 100
Dharma, 180
Dissociative identity disorder (DID), 94
DNA (Deoxyribonucleic acid), 16, 105-115,
Double Slit Experiment, 53-57, 85
Doyle, Sir Arthur Conan, 157
Dreamtime, The, 185
Dubois, Allison, 164
Dunne, Brenda, 135
Durkheim, Émile, 193-201
Dürr, Hans-Peter, 41
Dutch Study, 149
Eagleworks, 52
Eccles, John, 85-87
Eddington, Sir Arthur, 40, 85
Edward, John, 164
EEGs, 105, 131, 151
Einstein, Albert, v, 10-39, 44-49, 61-62, 100-118, 142, 147, 171, 177, 200, 205, 206, 214, 215, 217, 220, 222, 223, 238
Eippert, Falk, 92
Electromagnetic force, 38, 197
Ellison, Arthur, 162-163
EmDrive, 29
Emergent properties, 42, 97
Entanglement, 30, 44-46, 69, 99, 110, 111, 131, 230
Epigenetics, 109
Epilepsy, 95
Epistemological, 10
Erzberger, 9

Euclidean geometry, 28
Evolution, 10, 16, 19, 31, 39, 53, 66, 101-121, 165, 171-177, 209, 219, 221, 223, 225, 229, 237
Ewald, Professor, 192
Existential Anxieties, 171
Explicate Order, 230
Fenwick, Peter, 151
Fetta, Guiddo, 29
Feyerabend, Paul, 218
Feynman, Richard, 56
File-drawer effect, 143
Filippenko, Aleksey, 14
Finniss, Damien, 92 - 94
First cause, 68, 199
Flynn Effect, 99
fMRI, 89, 92, 131, 146
Foy, Robin and Sandra, 160-162
Frederick, Shane, 24-26
Free will, 80, 86, 89, 136, 174, 178, 201
Friesike, Sascha, 219
Gabor, Dennis, 226
Galileo, 30, 32, 118, 177, 195, 223
Galison, Peter, 49
Gall, Franz Joseph, 80
Ganzfeld Test, 129
Gauld, Alan, 160
Global Consciousness Project, 123-124
God particle, 197
Godel, Kurt, 51
Goldilocks Zone, 63
Goodwin, Brian, 108
Goswami, Dr Amit, 193,
Gravity, 29, 31, 37, 38, 39, 47, 63, 79, 83, 102, 148, 191
Greene, Brian, 47-48
Greyson, Bruce, 151
Gurwitsch, Alexander, 109
Haldane, J.B.S., 10
Halpern, David, 26, 27
Hameroff, Stuart, 101, 199, 208, 215

Hancock, Graham, 31
Harris, Bertha, 157
Hawking, Stephen, 28, 29, 39, 40, 42, 43, 50, 51, 52, 61, 63, 82, 84, 85, 97, 199, 214, 222
Heisenberg, Werner, 61, 84, 169
Higgs Boson, 197-198
Hinduism, 178, 180, 186
Hodgson, Richard, 158
Hologram, 43, 226-231
Holographic Universe, 209, 215, 226
Howard, George, 130
Hugo, Victor, 99
Hyslop, James, 158
Implicate Order, 185, 211, 215, 230
Inedia, 188-191
Infinite monkey theorem, 68
Institute for Venture Science, 217, 238
Institute of Noetic Sciences, 139, 161
Intelligent Design, 116-119
Interference patterns, 55, 229, 230
Intergovernmental Panel on Climate Change, 127
Interval Research Corporation, 140
Irreducible complexity, 116
Islam, 118, 170, 171, 178, 184, 188, 223
Islamic Golden Age, 171
Israelsson, Ulf, 161
Isvara, 186
Jahn, Robert, 135, 139, 212, 214
James, William, 157, 159, 170
Jani, Prahlad, 188
Jefferson, Thomas, 18
Jemison, Mae, 52
Jesuits, 195
Jesus, vii, 90, 91, 183, 189, 190, 191, 222, 223
Jones III, John E, 119
Jordan, Pascual, 62
Judaism, 178, 182-188
Jung, Carl, 220
Kahneman, Daniel, 24, 25, 26,

Kahuna Hawaiian healers, 131
Karma, 180
Kelle, 9
Kelvin, Lord, 13
Kim, Yoon-Ho, 56, 60,
King George of Greece, 157
Kirlian photography, 113
Kuhn, Thomas, 10, 21, 23
Laibson, David, 26
L'Aigle, 17
Laozi, 181
Large Hadron Collider (LHC), 197
Lashley, Karl, 228
Lavoisier, Antoine, 17
Layering, 9, 223
Lederberg, Joshua, 109
Leonid Meteor shower, 18
Levitation, 160, 187
Libet, Benjamin, 105,
Lincoln, Abraham, 157
Linde, Andrei, 199, 200
Lingelbach, Bernd, 22, 23
Lisieux, Therese, 189
Lodge, Sir Oliver, 158, 159
Lommel, Pim van, 109, 112, 148-150
Lorber, John, 104
Lost on the Moon, 11
Lucas, George, 5, 207, 214, 226
Malfoy, Lucius, 177
Martin, John, 114
Marx, Karl, 193
Maslow, Abraham, iii,
McCartney, Paul, 77
McCoy, Terrence, 240
Measurement problem, 55, 135
Meditators, 131-138, 189, 207
Medium (spiritual), 153-166, 223, 239
Mega-phenomena & Mega-phenomenon, iv, 210, 213, 224, 225, 242
Mendel, Gregor, 177
Mengele, Josef, 118
Meteors, 16, 63

Michelangelo, di Lodovico, vi, 102, 103
Michelson, Albert, 13
Microtubules, 101
Miller, Scott, 95-96
Mlodinow, Leonard, 43
M'Naghten, Daniel, 114
Moksha, 180
Morgan, Sally, 164
Morphic resonance, 99, 109
Morphogenetic field, 109
Morris, Robert, 160
Morton, John, 111
Mossbridge, Julia, 141
Mott, Sir Nevill, 176
Mozart, Wolfgang Amadeus, 77, 157, 229
Mullins, Kary, 140
Myers, PZ, 31
Naber, Father, 189
NASA, 5, 11, 30, 40, 52, 88, 132, 134, 135, 161, 189, 214, 217,
Nasr, Seyyed Hossein, 184
Natural laws, 28, 64, 78, 79, 82, 83, 97, 119
Naturalist views, 78, 79, 81, 82, 94, 119, 157, 165, 206, 246
Near Death Experience, 101, 148-150
Nelson, Roger, 123
Neumann, John von, 62, 135
Neumann, Therese, 189-192
Newman, Louise, 174
Newton, Sir Isaac, 42, 53, 100, 119, 171, 177, 217, 218
Non-local behaviour, 44, 45, 99, 112, 223
Norton, Mike, 26
Nudge, 26-27
Obi wan Kenobi, 226
On the Origin of Species, vi, 105, 116
Ontological, 10
Opal, 33
Paley, William, 117

Pallas, Peter, 18
Parnia, Sam, 151
Patañjali, 186, 188, 189
Pauli, Wolfgang, 62
Pearsall, Paul, 113
Peel, Robert, 114
Penrose, Sir Roger, 87, 100, 101, 105, 208, 212, 215
Persinger, Michael, 152
Phelan, Laurence, 240
Phrenology, 80
Piper, Leonora, 157-159
Placebo Effect, 90-96, 165, 207, 215, 230
Planck, Max, 21, 61, 62, 118, 147, 172, 177, 192, 199, 213, 217, 223
Plato, 4, 5, 14, 98, 100, 212, 232
Podolsky, Boris, 44
Polkinghorne, Sir John, 66, 85, 99, 224
Pollack, Gerald, 217
Pool table full of balls, 86-87
Precognition, 127, 146, 187
Presentment, 140-146
Pribram, Karl, 214, 228-230
Price, Pat, 134
Princeton Engineering Anomalies Research (PEAR), 135
Probability, 127
Prozac, 91
Pseudo-science, 16,
Psi, 126-130, 140-148
Psychokinesis, 127, 135, 139, 187, 207
Puthoff, Hal, 132
Putnam, Robert, 173
Pyramids, 4, 78
Quantum Measurement Problem, 135
Quantum Mechanics, 14, 39, 45, 62, 97, 98, 101, 111, 142, 165, 206, 208
Quantum physics, vi, 16, 21, 30, 40, 42, 61, 64, 66, 68, 78, 82, 84, 85, 100, 106, 149, 165, 199, 206, 218, 225, 245
Quantum State, 66

Quarks, 40
Qur'an, 31
Radin, Dean, 122, 130, 138, 187, 214
Randall, Lisa, 38-39
Random Event Generator (REG), 135
Rational choice theory, 26, 120
Readiness Potential, 105
Redesign My Brain, 22
Reductionism, 85, 184,
Remote Viewing, 132-134, 219, 230
Rosen, Nathan, 44
Rumsfeld, Donald, 15
Russell, Bertrand, 18, 79, 80, 175
Samkhya, Philosophy, 186
Samsāra, 180
Sartori, Penny, 151
Schaer, Hans, 161-162
Schlitz, Marilyn, 139, 161
Schmidt, Brian, 40, 46
Schmidt, Helmut, 138
Schrödinger, Erwin, 12, 44, 61
Schwartz, Gary, 163
Schwartz, Stephan, 219-220
Scientific Method, vi, 76, 84, 118, 126, 157, 171
Scole Experiments, 160-163
Second law of thermodynamics, 88
Seinfeld, Jerry, 77
Shah, Dr Sudhir, 188
Shawyer, Robert, 29
Sheldrake, Rupert, 29, 31, 82, 99, 109, 122, 151, 177, 212, 214, 218, 222, 228
Sherrington, Charles, 87
Siddhis, 187-188
Silliman, Benjamin, 18
Sinclair, Upton, 21, 147
Six Blind Men and the Elephant, iv, 6, 33
Snake Oil Salesmen, 153
Snyder, Allan, 23, 239
Society for Psychical Research (SPR), 157, 160

Socrates, 14
Souls, 166, 177, 185-186
Space-time, 5, 49-52, 61, 66, 69, 81-104, 115, 112, 122-125, 138, 142, 165, 166, 176, 199-207, 215, 225
Spemann, Hans, 109
Spielberg, Steven, 48
Squires, Euan, 62, 136
SRI International, 122, 130, 140
St. Francis, 190
Stalin, Joseph, 98
Stanford Research Institute (SRI), 132
Stanovich, Keith, 24
Stapp, Henry, 62, 85
Star Trek, 4, 5, 51, 52, 134, 210, 214, 216, 231
Star Wars, 4, 5, 77, 207, 214, 216, 226
Statistically significant, 127, 135
Stem cell, 109
Stigmata, 189, 190, 191
Streby, Henry, 144-145
String Theory, 39, 50, 51
Strong nuclear force, 38
Strong View, 62, 135
Subjectivity, 7, 56, 59, 60, 61, 84, 87, 107, 151, 169, 179, 195, 196, 232
Subspace, iv, 120 and through document
Sunstein, Cass, 26
Super-organisms, 42
Superposition, 56, 68, 101, 112, 199
Surowiecki, James, 10
Survival instinct, 106, 107, 175
Survival of the fittest, 107, 115, 117, 174
Susskind, Leonard, 43
Sūtras, 186, 187,
Swift, Jonathan, 148
Sylvia, Claire, 113
Tacey, David, 174
Tait, Tim, 41
Taoism, 178, 181
Targ, Russell, 132, 133, 134, 189

Tart, Charles, 143
TEDx, 31
Tegmark, Max, 48, 64, 65
Telepathy, 127, 128, 146, 165, 187,
Ten-Choice Trainer, 143
Thaler, Richard, 26
The Lord Of The Rings, 23
The Traveller, 210
Theory of Everything, 50, 224, 231
Theory of Relativity, 14, 39, 46, 47, 51, 100, 214, 215
Thurston, Herbert, 190
Tiller, William, 231
Time vortex, 4
Time Warps, 51, 214
Top-down effect, 84, 86,
Tressoldi, Patrizio, 141
Treynor, Jack, 10
Triangulation, iv, 6-12, 32, 34, 178, 208, 223, 238
TRIZ (in English "the theory of inventive problem solving" or TIPS), 213, 226
Tsemberis, Sam, 240-241
Tunnelling, 112
Turin, Luca, 110, 112
Tversky, Amos, 24
Tymn, Michael, 159
Tyrannosaurus, watch wearing, 64
Ullman, Montague, 220
US National Academy of Sciences, 176, 222
Utts, Jessica, 141

Values, 114-116, 173, 174, 175, 193, 200, 201, 207, 208, 211, 237
Vatican Observatory, 221
Victoria, Queen, 157
Voyager, 3
Wallace, Alfred Russel, 19, 20, 105, 106, 117, 119, 121, 153, 157,
Wallace, David Foster, 208
Warp Fields, 210
Weak nuclear force, 38
Weber, Max, 193
Weiss, Paul, 109
Wells, H. G., 214, 215
West, Donald, 160
West, Richard, 24-25
Wheeler, John, 66, 69, 199
White, Harold 'Sonny', 52
Whittington, Mark, 30
Who Wants to Be a Millionaire, 11
Wigner, Eugene, 62, 85, 98
Wilber, Ken, 231-238
Wildey, Chester, 145
Witch trials, 154
Witches, 153-156
Witten, Edward, 50-51
Wittgenstein, Ludwig, 216
Yoda, 207
Yoga, 180, 186-189
Yogas, 180
Young, Thomas, 53-57
Zeilinger, Anton, 68, 144
Zimmerman, Michael, 221, 22
Zwartz, Barney, 174

www.ingramcontent.com/pod-product-compliance
Lightning Source LLC
Chambersburg PA
CBHW021119300426
44113CB00006B/206